Engineering Construction Materials

KHALID HASSAN SIDDIQUI
B.Sc (Hons), M.Sc (Geology)

FEROZSONS (Pvt.) LTD.
LAHORE - RAWALPINDI - KARACHI

ISBN 978-969-0-02728-3

First published 2021 by
Ferozsons (Pvt.) Ltd.
81-D/1, Main Boulevard Gulberg-III, Lahore
277, Peshawar Road, Rawalpindi
Mehran Heights, Main Clifton Road, Karachi

Siddiqui, Khalid Hassan

Engineering Construction Materials

Published by
Zaheer Salam, Ferozsons (Pvt.) Ltd.,
81-D/1, Main Boulevard, Gulberg-III,
Lahore-54000, Pakistan

Printed in Pakistan at
Ferozsons (Pvt.) Ltd., (Printing Div.) Lahore.

email: support@ferozsons.com.pk
www.ferozsons.com.pk

Contents

Acknowledgement

My sincere and heartfelt gratitude to National Engineering Services of Pakistan under whose guidance I had the opportunity to work on various construction projects in Pakistan and gained knowledge and technical information of new technology and research on materials.

I have sincerely tried to write what I have gained in knowledge and experience from my work, books and technical publications. I am indebted to various authors whose work has helped me in writing this book.

I wish to express my thanks to all my colleagues and friends with whom I worked on various Highway and building projects and achieved valuable technical information through detailed discussions and exchanging views with each other.

I would also like to express my gratitude to Asphalt, American concrete Institute, all volumes of AASHTO and ASTM standards whose valuable literature and materials specification were of great help in writing this book.

Introduction

The topics covered in the book are mostly related to the material, which is the most important part of Highway Engineering. The topics have been discussed in essays, keeping in view the increasing importance of Highway material and its major contribution in civil and Highway engineering. All the technical problems related to highway construction have been discussed in view of present techniques and research on the subject. This book includes following topics, in which the prime importance and stress remains on the major contribution of materials in producing standard quality of roads and bridges.

1. Silty and clayey materials in Highway construction and industrial use.
2. Comments on WBM via aggregate base course.
3. Los Angeles abrasion test of large size coarse aggregate.
4. Why an improved subgrade is a better choice for durable and strong roads.
5. Tests on cores recovered from hardened concrete.
6. How to check the admixture used in concrete mix.
7. Load calculation of yield and ultimate strength of concrete pipes
8. Durability of concrete.
9. Crack filling in concrete structure.
10. A look at concrete sealers.
11. Hauling time of concrete mix needs to be extended.
12. Pavement construction – rigid or flexible – merits and demerits.
13. Grading of aggregates and workability of concrete aggregate.
14. Visual inspection and petrographic analysis of concrete aggregate.

15. Concrete in girders and grouting of bonded steel.
16. How to avoid failure in concrete construction.
17. Concrete mix design class A-2-constriction of barriers with machine.
18. Water proofing–safety and protection for concrete.
19. How to achieve good riding quality of pavement.
20. Standard asphalt paving equipments and temperature of mix in construction of asphalt pavements.
21. Asphalt Emulsions.
22. Chemical requirements–Deformed and plain billet steel bars for concrete reinforcement.
23. Expansion joints.
24. Need to review the project specification.
25. PVC pressure pipes.
26. Quality control of bearing pads and comparison before aging and after aging.
27. Water stops in concrete construction.
28. Non destructive tests on concrete structures.
29. Drying shrinkages of concrete.
30. Low alkali cements.

The construction work at site needs the attention of contractor, consultant and client. If these three components are honest, devoted and sincere to the work, best output and results can be achieved. However most construction work of various projects raises these questions again and again.

1. Why our construction works do not come up to the standard.
2. Why huge amounts are spent on construction without gaining specified results.
3. Why the construction projects are awarded on lower costs when the specified quality remains unachieved.
4. Why the contracts are awarded to contractors who have none or little construction experience, and are not well equipped with the Highway machinery required at site. Have no laboratory equipment or qualified personnel.

5. Why is it that many construction projects are required to be completed within a short period of time even before their scheduled time, in order to show efficiency to the superiors.

Such question always remain unanswered for various reasons. However honesty and devotion should be the prime concern of everyone involved with the project specially the high officials and site supervision staff.

I have tried to make the book simple so more people can understand it.

Preface

The book embraces various topics concerning concrete and properties related to other construction materials including cement and aggregate. Topics related to soil and asphalt material have also been covered in the present publication keeping in view the importance of such materials in Highway and building construction. The book is in essay form and provides the solutions to all problems which a construction Engineer faces from time to time while working at site. Concrete is in ever widening use and is the main pillar of construction Industry. In addition to dealing with fundamentals of materials, specialized knowledge concerning concrete, asphalt and soil has been added to the book. The book also provides the latest scientific and technical information and the trend of research on individual problem. All precautionary measures necessary to be adopted during the process of concrete pouring at site and operation of concrete batching plant have also been added to the present publication. Because of the multitude of problems faced during construction process, the quality control of concerned materials at site has been taken into consideration in order to achieve best and durable concrete structures and asphalt pavements.

The present publication also provides detailed knowledge about formation of rigid pavements and how such concrete pavements differ from flexible pavements keeping in view the merits and demerits of such structure. Whatever has been written in the book is based upon knowledge collected from various technical research papers, books of various authors and some bulletins related to the subject. All references used in writing this publication have been given at the end of the book. The technical knowledge in the book comes from the author's own experience while working on various projects with contractors, consultants and clients.

This is a small effort and would act as a dim light in the dark. I pray to Almighty Allah to help me and give me courage to serve the mankind and my beloved country Pakistan.

Use of Low Alkali Cements

Quality Control on Cement Production

In ordinary Portland cement (OPC) the specified quantity of alkali contents is up to 1% but generally the alkali contents are about 0.76% Low alkali cements are those cements in which percentage of alkali contents are less than 0.60%.

Formula for calculating the alkali contents is $Na_20+0.658\ K_2o$. Generally the OPC cements contain the following range of N_2o and K_2o:

Na_2O	0.252—0.234
K_2O	0.743—0.550

Which indicate the alkali content 0.74% to 0.596.

When sodium and potassium oxide in excess quantity in cement react with certain siliceous component of some aggregate, such as opal and tridymite they form a calcium alkali silica gel which causes expansion. This expansion causes disruption and disintegration of the concrete. Some carbonates have also been known to produce a similar effect. This problem may be overcome by the use of cements of low alkali content or possibly by incorporating a proportion of ground granulated blast furnace slag. In low alkali cements special clays like china clay, bentonite and flint stone are used. The limestone used for low alkali cements is generally pure limestone containing minor impurities, in some cases limestone which contains no impurity. The special clays used in the cements have low alkali contents.

Alkali silica reaction can also be controlled by using supplementary cementation materials in proper proportion. Silica fumes, fly ash and ground granulated blast furnace slag have significantly reduced expansion due to alkali silica reactivity. In addition compounds have also been used to reduce ASR. It is important to note that not all ASR gel reactions produce destructive swelling. Sulphur in excess also causes build up and the ratio of alkali to sulphate called degree of sulphatisation must be kept between 0.8 to 1.0% Excess sulphur limits the addition of gypsum in cement which may cause setting problem.

Typically equi dimensional pattern cracking is characteristic of non-reinforced or lightly reinforced concrete affected by alkali aggregate reaction, when there is significant amount of re-enforcing steel tends to be more prominent in the direction parallel to the re-enforcement. However map cracking can also result from other causes such a drying shrinkage (plastic shrinkage cracking). Concrete deterioration caused by alkali aggregate reaction is generally slow, but progressive.

The limestone used in low alkali cements should be pure limestone containing minor impurities or no impurities. Therefore while selecting raw materials, the problem of impurities should be taken into consideration.

1. Free silica if present in raw materials, causes high wear in the crusher and in the grinding mill. Raw materials containing silica must be ground very finely. It is also necessary that raw material containing more impurities should be avoided. Raw materials containing phosphate, lead and zinc or their compound should be avoided.

Chemical analysis%

Slo_2	Al_2o_3	Fe_2o_3	Cao	Mgo	K_2o	Na_2o	So_3
20-4	4.75	3.89	62.4	2.78	0.5	0.20	2.2

Magnesium if present in excess amount in clinker causes long term unsoundness in cement produced and hence needs to be restricted to 5% in almost all the standard specification for Portland cement.

Limestone and clay are the major raw materials which need a detailed geological exploration. About 1500 – 1600 kg of the average amount of materials are required to produce one tonne of Portland cement. The average amount of materials required to produce one ton of Portland cement are $^1/_2$ to $^1/_3$ ton of calcareous material like limestone and marl and 0.3 – 0.4 ton of argilaceous material like clay and shale. Also 0.05 to 0.06 ton of natural or synthetic gypsum is required per ton of Portland cement. Typically 75 to 80% calcareous material is present in the kiln feed, remaining, being angilaceous and other additives. Raw materials required for low alkali cements should be of adequate purity and uniform quality.

Calcareous materials like limestone, marl, chalk or sea shells provide the limestone component whereas clay, shale, sand, slate and pozzolans provide silica, alumina and iron components.

A typical composition of general purpose (ordinary) Portland cement clinker is 22% silica, 4.5% alumina, 3.5% iron oxide and 65% lime thus totalling to 95%. The balance is 1% free lime and 4% of magnesium, surplus, sodium, potassium and other miner oxides.

ASTM C – 219 – 98 – Composition of phases/minerals.

1.	Tricalcium silicates	Alite	C_3S	26.32% Slo_2 73.68% Cao
2.	Dicalcium silicate	Belite	C_2S	34.89% Slo_2 65.11% Cao
3.	Tricalcium aluminate	–	C_3A	37.74 Al_2o_3
4.	Tetracalcium aluminoferite	Celite	C_4AF	20.98 Al_2o_3

Alite and belite comprises the silicate phase. Since about 65% of OPC cement clinker consists of calcium oxide, so cement manufacturing plants are usually situated near the deposits of limestone's and sometimes of clay or shale.

Alkalies (sodium and potassium) if present in excess quantities, result in build up in the kilns and preheaters, thus affecting the kiln operation. If present more then 0.60% as sodium oxide equivalent

(% N + 0.658% K) causes alkali aggregate reactions and damage concrete structures as already discussed.

Silicate phases are dominating and are responsible for the strength development of cement. Higher contents of alite in clinker and fine grinding results in rapid hardening cement. This phase ranges from 40 to 70% in Portland cement clinker.

During the production of cement its fineness by sieve No.200 should be kept in view. The object is to check the proper grinding of the cement. The cement must be ground during manufacture to be uniformly fine, otherwise the concrete made will be of poor quality and require a large amount of mixing water. Bleeding also occurs that is even before the concrete begins to set initially, water will come out to the top due to the settlement of solids. The fineness of cement should not be less them 2850cm. sq/gm (ASTM C-150). The auto clave expansion (soundness) of cement should not exceed 0.8% (ASTM C-150). Initial setting time of cement should not be less than 45 minutes whereas its final setting should not be more than 375 minutes (ASTM C-150). Three (3) days and seven (7) days compressive strength of cement should not be less then 1800 psi and 2800 psi respectively (ASTM C-150).

During the production of cement very definite proportions of the raw materials must be maintained as the resulting properties of the cement are affected by difference of raw material or proportions. By increasing the lime percentage with respect to silica for instance, the rate of gain in strength is increased and heat liberation during the process is increased also.

Drying Shrinkage of Concrete

Cement is the most important and most expensive ingredient in concrete. It is the basis of action of most of the admixtures. Cement is also the biggest problem in concrete–it requires the most water and gives off the most heat. The strength of any concrete at any age depends on the degree to which the originally water filled space becomes a smaller fraction of the volume of concrete. Therefore if we can reduce the total free water in the concrete system, we will reduce shrinkage and increase strength with the judicious use of admixtures and cement. Water cement ratio limitations tend to increase cement contents beyond what is needed for the most effective use of today's admixtures. The more cement in a mix, means the more total shrinkage. Compressive strength of the cement needs to be tested before it is used at site. Laboratory mix design, with admixtures become increasingly important in determining the correct mix for the job. It is unprofessional to give a strength mix design to a contractor with a water cement ratio limitation that does not allow him to place it properly and is in reality a mix of higher compressive strength. Workable concrete mixes are always appreciated with high compressive strengths. Different cement used for the same intrinsic quality of concrete will have different water demands at the same workable slump.

There are no water cement ratio guide lines for mixes containing high range water reducers, fly ash, micro silica etc to obtain the properties such as strength, durability and impermeability that designers seek. Therefore the best measure of water content of any mix is the slump of that mix. It is the use of plasticizers and super plasticizers that make the concrete workable and even reduce the heat of hydration. The concrete mix should contain a balanced amount of cement, neither more or less and water cement ratio should also be within the specified range depending upon the various classes of concrete. Minimum shrinkage in concrete can only be achieved by using a balanced mix where the quantity of cements and water cement ratios are well controlled for various classes of concrete. This is the main reason why many specifications apply limits to the maximum cement content to pre-stressed concrete to reduce excessive long-term movements such as shrinkage and creep. However cements with the lowest water demand are typically those blended with fly ash such as BS 6588 with 15 to 35% fly ash and BS-6610 with 35 to 50% fly ash, where at least a 15 lit m3 decrease in water demand may be expected against ordinary Portland type I cement. Rapid hardening or type III Portland cement require generally 5 to 10 liter more. It is therefore advisable to protect the materials from solar heat gain and to water spray the aggregates to keep them cool by evaporation. It can thus be more easily understood why there have been so many durability failures in the Middle East where questionable waters were used in combination with very hot conditions which inevitably caused considerable increase in water demand. It is therefore necessary to minimize shrinkage and curling and high range water reducers /admixtures can play an important role in this case- slump no longer means low water content. It may simply mean use of water reducing admixtures. The proper way to achieve impermeability is by lowering the water content by use of super plasticizers. The drying shrinkage of concrete should not exceed 0.045% (ASTM C-596-81) and the initial drying shrinkage of concrete should not exceed 0.06% (BS-1881).

Performance of fresh concrete in terms of finish ability and plastic shrinkage cracking is a matter of quality of silica fume, additive used, curing techniques and availability of equipment and skilled labour. All forms of silica fume tend to decrease chloride permeability of concrete. There is evidence that the more impermeable the concrete, the greater its durability and there is less chance of harmful chloride ions getting to the steel. Permeability is a function of density and lack of capillaries and cracks.

Plastic shrinkage cracks occur mostly in horizontal surfaces, usually develop at the time water sheen disappears from the surface of the concrete. They are more apt to occur during hot dry weather rather than moist or cool weather. Excessive rapid evaporation of water from the surface of concrete specially in dry weather is the major cause of plastic shrinkage cracking. Rate of evaporation depends on air temperature, relative humidity, concrete surface temperature and wind velocity at the concrete surface. To avoid drying shrinkage cracks in concrete the following precautionary measures should be adopted.

A. Lower the temperature of concrete during hot weather.
   1. By cooling mixing water.
   2. Shading and spraying aggregate with water.
   3. Shading forms and /or subgrade.
   4. Placing concrete during cooler period of the day.

B. Increase humidity at concrete surface by
   1. Fog spaying during finishing operation until curing can be started.
   2. Protecting the concrete with temporary coverings during any delay between placing and finishing.

Plastic shrinkage cracks can be eliminated in some cases by applying curing compound immediately where hard troweling is not required. Curing compound to be applied after initial set has occurred.

Hardened concrete will shrink about 0.05% of its length when it dries. It will shrink more if the concrete when mixed, contained a large amount of water (due to high slump, small maximum size of aggregate, angular shape of aggregate, excessive amount of fines or poor grading). Also, the aggregate itself, by means of the bond of paste to the aggregate particle and the modulus of elasticity of the aggregate affect shrinkage of the concrete. The shrinkage of the paste is resisted by dense stiff aggregates that bond well to the paste. At a shrinkage of 0.05% concrete will shrink almost $1/16^{in}$ (1-6 mm) in 3 meter of length and if the concrete is restrained it will crack. Concrete placed at hot noon day temperature will contract as it cools at night. It is not uncommon to have a drop of 22C° between the day and night in some desert areas. Such a drop in temperature would cause the concrete to contract 0.025 inch (0.6 mm) in 3 meter in length.

Cracking due to shrinkage and temperature changer can be controlled to a large extent by steel reinforcement. Smaller sized bars spaced close together are more effective in controlling cracking than same percentage of steel using large diameter bars spaced relatively for apart. Cracking can also be controlled by per-stressing or post-tensioning the concrete to keep it in compression so that tensile stresses do not develop.

Use of steel concrete fibre (needle type shape) in concrete mix are useful to increase the compressive strength of concrete and also to reduce shrinkage factor in concrete.

Methods of Placement of Concrete

Precautionary Measures in Concrete Pouring – Concrete Under Water

Placement methods:

1. By Bucket
2. By Pumping
3. By Chute
4. Direct discharge from truck.

Concrete should be placed so as to avoid segregation of the materials and the displacement of the reinforcement. Loss of fines or any other way which may cast a deteriorating effect on the concrete quality should be avoided.

Open troughs and chutes should be of metal a meter lines where steep slopes are required. All chutes, troughs and pipes should to kept clean and free from coatings of hardened concrete by thoroughly flushing with water after each run. Water used for flushing should be discharged. In case of pumping operation height of drop of concrete should remain 100cm to 150cm. However 100cm drop height would be preferable. Rate of pumping should be 100-150 cubic meter per hour and the slump should range from 100-160 mm in case admixtures are used. When placing operations would involve dropping the concrete more than 1½ meters, it should be conveyed through sheet metal or other approved pipes as far as practiceable, the pipes should be kept buried in the newly placed concrete.

Pumping

The operation of the pump should be such that a continuous stream of concrete without air pockets is obtained. When pumping is completed, the concrete remaining in the pipe line, if it is to be used, should be ejected in such a manner that there will be no contamination of the concrete or separation of ingredients. After the entire operation, the equipment should be thoroughly cleaned.

By Bucket

When concrete is placed with a bottom dump bucket, the top of the bucket should be open. The bottom doors should open freely down ward and out ward when tripped. The bucket should be completely filled and slowly lowered to avoid backwash. It should not be dumped until it rests on the surface upon which the concrete is deposited and when discharged should be withdrawn slowly until well above the concrete.

Compaction

Concrete during and immediately after placing should be thoroughly compacted except lean concrete under footings and concrete deposited under water. Concrete in walls, beams, columns etc, should be placed in horizontal layers not more then 30 cm thick Each layer should be placed and compacted before the preceding layer has taken initial set to prevent injury to the concrete and avoid surfaces separation between the layers.

Each layer should be compacted so as to avoid the formation of a construction joint with a preceding layer, which has not taken an initial set. The compaction should be done by mechanical vibration. The concrete should be vibrated internally unless special authorization of other methods is given by the Engineer. The intensity of vibration should be such as visibly to affect a mass of concrete with a 5cm slump over a radius of at least half a meter. A sufficient numbers of vibrators should be provided to properly compact each batch immediately after it is placed in the forms. Vibrators should be manipulated so as to thoroughly work the concrete around the reinforcement, embedded fixtures, into the corners, angler of the forms and should be applied at the point of placing and in the area of freshly placed concrete. The vibrators should be inserted in to and withdrawn from the concrete slowly. The vibration should be of sufficient duration and intensity to compact the concrete thoroughly. Application of vibrator should be at points uniformly placed and not farther apart then twice the radius over which the vibration is visibly effective.

Vibration should not be applied directly to the reinforcement or to the sections as Layers of concrete that have hardened to the degree that concrete ceases to be plastic under vibration.

It should not be used to make concrete flow in the forms over distances so great as to cause segregation. Vibration should be supplemented by such spading as is necessary to insure smooth surfaces and dense concrete along form surfaces and in corners and locations impossible to reach with the vibrators.

Internal vibrator with immersed tube may be used and the frequency of the surface vibrator should not be less then 3500 vibrations for minute.

Placing Concrete Under Water

Concrete to be deposited under water should be class A-2 or A-3 with a minimum cement content of 400 kg per cubic meter. The slump of the concrete should be maintained between 150 to 170 mm. To prevent segregation, it should be corporeally placed in a compact mass in its final position by means of a trimie and it should not be disturbed after being placed. Water must not be allowed to flow past the fresh concrete surface.

A trimie should consist of a tube having a diameter of not then 25cm constructed in sections having flanged coupling fitted with gaskets with a hopper at the top. The trimie should be supported so as to permit free movement of the discharge and over the entire top surface of the work and so as to permit rapid lowering when necessary to retard or stop the flow of concrete. The discharge end should be closed at the start of the work so as to prevent water entering the tube and should be completely submerged in concrete at all times. The trimie tube should be kept full to the bottom of the hopper. When a batch is dumped into the hopper, the flow of concrete should be induced by slightly raising the discharge end but always keeping it in the placed concrete. The flow should be continuous until the work is completed.

When the concrete is placed with a bottom dump bucket, the top of the bucket should be open. The bottom doors should open freely downward and outward when tripped. The bucket should be completely filled and slowly lowered to avoid backwash. It should not be dumped until it rests on the surface upon which the concrete is to be deposited and when discharged should be with drawn slowly until above the concrete.

Dewatering may proceed when the concrete seal is sufficiently hard and strong. Though concrete presents an image of stability and permanence, it is actually a fragile material that demands careful handling.

Testing of Cores Recovered from Hardened Concrete

It should be noted that according to ACI 318-89 (Revised 1992) cores are tested in dry state.

Average strength of three (3) cores should be equal to at lest 85% of specified strength and no single core should be less then 75% of specified value.

NEVILLE Testing of hardened concrete:

Specimens containing embedded reinforcement should not be used for determining compressive strength, splitting tensile strength or flexural strength. T 24-2 5.1.2. Sampling AASHTO.

Historically it has been assumed that core strengths are generally 85% of the corresponding standard cured cylinder strength but this is not applicable to all structures, the acceptance criteria for core strength is to be established by the specifying authority. T 24-2, 3.5.

In case of embedded steel, cores give less strength and the reduction ranges between 8 to 18% – the reduction seems to be higher when the height/diameter ratio of the cores is double than lower value of this ratio. NEVILLE Testing of hardened concrete.

How to test the concrete cores

Submerge the test specimen in lime saturated water at 23 $\pm$ 1°C for at least 40 hours immediately prior to making the compression test. Test the specimen promptly after removal from water storage. During the period between removal from water storage and testing, keep the specimen moist by covering with a wet blanket of burlap or other suitable absorbent fiber.

AASHO T-24 (ASTM C-42)
ACI 301 chapter 17.
ACI 318 chapter 5.

Before making the compression test, cap the ends of specimen in conformance with the procedure prescribed in T-231. Prior to testing measure the length of the capped specimen to the nearest 2.5 mm and use this length to compare the length diameter ratio.

12.114 ASZY PUZA and J.S. Grossman
Cylinder vs core strength. Concrete International
12 No. 2 PP 55-61 (1990)

The location in the structure from which the core has been taken may affect the strength of the core. It the core has been taken from the concrete in tension, the core strength may be low because of the presence of cracks, thus a false picture, of the strength of concrete in the structure will be obtained.

The position of the core with respect to height of lift may also be of relevance. Cores usually have the lowest strength near the top surface of the structure.

NEVILLE Testing of hardened concrete:

When preparing strength test specimens from samples of hardened concrete, samples that show abnormal defects and samples that have been damaged in process of removal shall be used.

Fires, accidents or premature removal of form work always resulting in deficient strength.

ACI 301 1.7. acceptance of structures.

If it is possible to eliminate the reinforcement an L/D of 1.00 or greater can be used.

Core Drilling: A core specimen taken perpendicular to a horizontal surface shall be located, when possible, so that its axis is perpendicular to the bed of the concrete as originally placed. A core specimen for the determination of length shall have a diameter of at least 100mm (4 in). Measure the length of cores in accordance with the provision of T-148.

The diameter of core specimens for the determination of compressive strength should be at least 3 times the nominal maximum size of the coarse aggregate used in the concrete and must be at least twice the maximum nominal size of the coarse aggregate in the core sample. The length of the specimen, when capped, shall be as nearly as possible twice its diameter. A core having a maximum height of less than 95% of its diameter before capping or a height less then its diameter after capping shall not be used. The ends of core specimen to be tested in compression shall be essentially smooth, perpendicular to the longitudinal axis, and of the same diameter as the body of the specimen. It necessary, saw or tool the specimen until the following requirements are met.

If the ratio of the length to diameter of the specimen is less then 1.94 apply correction factors shown in the following table.

Ratio of length of cylinder to diameter	Strength correction factor	
1.75	0.98	AASHTO Designation
1.50	0.96	T- 24 -97
1.25	0.93	ASTM Designation
1.00	0.87	C 42-94

If the ratio of length to diameter of the specimen exceeds 2.10 it shall be reduced in length to fall within the ratio limits of 1.94 to 2.10 Specimens within these ratios, require no correction.

Compressive Strength of Cylindrical Concrete Specimens

This test method consists of applying a compressive axial load to molded cylinders or cores at a rate which is within a prescribed range until failure occurs. The compressive strength of the specimen is calculated by dividing the maxim load attained during the test by the cross sectional area of the specimen.

The testing machine shall be of a type having sufficient capacity and capable of providing the specified rates of loading. Verification of calibration of the testing machine needs to be done every 12 months after the previous verification. Calibration of compressive strength machine should also be done at the original installation or relocation of the machine or whenever there is reason to doubt the accuracy of the results without regard to the time interval since the last verification. The compressive strength machine must be power operated and must apply the load continuously rather then intermittently and without shock.

The percentage of error for the loads within the proposed range of use of the testing machine shall not exceed $\pm$ 1.0% of the indicated load.

Type of compressive strength machines

1. Machine with graduated scale needle (Hydraulic)
2. Electronic digital machines.

If the load of a compression machine used in concrete testing is registered on a dial, the dial shall be provided with graduated scale. If the testing machine load is indicated in digital form on the numerical display, it must be large enough to be easily read. For hydraulically operated machine the load shall be applied at the rate of 20 to 50 psi/s.

The curing environments of cylindrical mould should be adopted according to C 192/C 192m-95 ASTM. The specimen needs to be removed from the mold 24±8 hours after casting and these cylinders should be cured in water tank at 23±2c until the moment of test. The lest specimen should have free water maintained on the entire surface area. Kindly note that in case the temperature of curing water is more than 25°C, high range test results of compressive strength would be achieved.

Abnormal compressive strength of concrete cylinders

The compressive strengths of concrete cylinders taken at site should not be more than the designed strengths. In case the compressive strengths of concrete cylinders is more than the designed strength, it indicates doubt and all precautionary measures need to be adopted to ensure the actual compressive strengths of concrete. Abnormal high strength of concrete cylinders can not be acceptable and it reflects the poor quality control from consulting supervision staff. There can be various reasons for obtaining abnormal compressive strength attitude.

1. Compressive strength machine is not properly calibrated and it indicates the defective attitude.
2. Poor quality control from consulting supervision staff and special cylindrical molds are being cast at the time of concrete pouring or after the concrete pouring.

It is always advisable to exercise a strong check at the time of crushing of cylinders on the lowest and highest reading of compressive strength machine on a daily and monthly basis. Standard deviation of concrete needs to be calculated and a complete record needs to be maintained on a fortnightly and monthly basis. In the absence of all these checks concrete quality control con not be established.

It is always in the interest of work to use the hydraulic machinc with gradated scale and needle. For hydraulic operated machine the load can be applied at specified rate of 20 to 50 psi/s which provides a good quality control during operation of such machine.

The temperature of water curing tank also plays an important role to provide high strengths which needs to be controlled.

Strict supervision needs to be exercised on concrete batching plant as this is the only primary source of producing concrete mix. Concrete batching plant needs to be properly calibrated on monthly basis with the dead weights which must be made available on concrete batching plant all the time.

Concrete mix design must be strictly followed and supervised as this practice can ensure the quality control of concrete mix. Kindly note that casting and crushing of concrete cylinders can also involve unfair means from contractor side which may cause adverse effects on the quality control of concrete mix.

Admixtures Used in Concrete Mix

Admixture is a material that is used in concrete mix in order to modify the concrete to improve the properties of the mix in the wet or hardened state. Admixtures are normally supplied in liquid form to introduce through automatic dispensing equipment by volume on cementitious weight. Cement admixtures are formulated products, designed to be added to concrete, mortar or grout mixes in small amounts (less then 5% by weight on cement. Cementitious material includes Portland cements, PFA, slag and silica fume.

There are several categories of admixture and their performance and uniformity are controlled by national standards. Admixtures can be divided in to the following categories or types.

1. Plasticisers (also called water reducers) these can be subdivided in to
   A. Normal plasticisers/water reducers
   B. Superplasticisers/High range water reducers.
2. Air Entrainer's
3. Retarders
4. Accelerators
5. Special purpose admixtures.

   Seven (7) of admixtures

   Type A: Water reducing admixtures

   Type B: Retarding admixtures

   Type C: Accelerating admixtures

   Type D: Water reducing and retarding admixtures

   Type E: Water retarding and accelerating admixtures

   Type F: Water retarding high range admixtures

   Type G: Water retarding high range and retarding admixtures

1. Admixtures containing chloride (as cl in excess of 0.5% by weight of admixture), Fluorides, sulfite and nitrate shall not be used. Admixtures that contain relatively large amounts of chloride

may accelerate corrosion of pre-stressing steel (calcium chloride– according to ASTM C-1152.)

2. Loss on ignition should not exceed 4%

i. **Residue by oven drying:** When dried the oven dried residues of the initial sample and of the subsequent sample shall be within a range of variation not greater them 4 percentage points (Non liquid Admixture).

ii. When dried as specified the oven dried residues of the initial sample and of subsequent sample shall be within a range of variation not greater than 5 percentage points.

3. Specific gravity (Liquid admixtures) 1.17 $\pm$ 0.01 kg/letter.

When tested the specific gravity of subsequent test sample shall not differ from the specific gravity of the initial sample by more then 10% of the difference between the specific gravity of the initial sample and that of reagent water at the same temperature.

Slump Criteria to Check the Admixture

Slump criteria

Slump of concrete mix containing admixture should be checked at the time of production. The specified dose of admixture proposed by the manufacturer should be used. Usually the middle point of the range of admixture given by the manufacture is adopted. The concrete mix containing admixture pertaining to class A-3 should provide slump 150 to 160 mm at the time of production from concrete batching plant. The dose of admixture is adjustable according to weather conditions which vary. Concrete class A-2 with admixture should provide the slump 150 to 140 mm. Class A-1 concrete containing admixture which is mostly used in box culverts and drains should provide slump ranging from 130 to 110 mm.

The slump of concrete with admixture must be noted at the time of production of concrete and checked every ½ hour for two hours. Downfall of slump of concrete mix should not be less than 10% after one (1) hour and 25 to 30% after two (2) hours.

Admixtures maintains durability in concrete mix. Durability is largely controlled by the permeability of the concrete and the best way to achieve low permeability is the use of water reducers which cut water cement ratio below 0.45 to 0.40.

Why use admixtures

Admixture can be used to improve the quality and cohesion of concrete. They can also be used to improve placing technique by achieving desired workability. However the major advantage of admixture is seen in terms of durability in modern concrete construction methods.

Slump retain test is the best one to judge the performance of admixture in concrete mix.

Cement – A Source of Plasticisor

Cement acts as a plasticisor but it is not a good substitute of actual plasticisor due to its cost compared to plasticisor. Cement is a very costly item, but not beneficial if used in concrete mix in high quantity. Maximum water cement ratio for class A-2 concrete is normally kept at 0.45. Cement content normally used is 450 kg/M3 to achieve specified strength which indicates that about 203 kg water would have to be used to achieve this water cement ratio. If the same water cement ratio is maintained without using admixture it would be maintained at the expense of excessive cement and water and in this case 520 kg cement/cum would have to be used with water content of about 234 kg to achieve the specified workability and strength. Use of excessive quantity of cement in place of admixture is incorrect and it may cause cracks and affect the durability of concrete.

It would be beneficial to use up to 450 kg per cu.m cement for concrete class A-2 where water cement ratio is 0.45 which is normally maintained. Water cement ratio for class A-1 concrete is 0.50 maximum and cement quantity is around 350 kg per cu.m. However class D-1 concrete which is mostly used in pre-Stressed girders, quantity of cement can be up to 510 to 530 kg/cu.m with water cement ratio 0.40 maximum.

Tests on Concrete Structure

Rebound hammer and ultrasonic pulse velocity measurements tests give a direct measure of compressive strength of concrete and such tests also give comparisons between suspect concrete and similar concrete that is known to be satisfactory.

Useful information can often be obtained by pounding the surface of concrete with an ordinary hammer. The difference in sound when the concrete is struck may identify areas of delamination of the concrete.

In the rebound hammer (Schmidt hammer) a mass strikes the concrete surface at a constant velocity and the distance through which it rebounds depends upon the hardness of concrete near the surface. Calibration curves are available that give an indication of the relationship between rebound and compressive strength. In order to get a reliable result, however, readings should first be taken on a cube of similar concrete that is then crushed in order to provide a reference point. A recording type of Schmidt hammer is available in which the rebound numbers are automatically recorded on a roll of paper. It must be remembered that the strength of concrete in a structure will not be the same as the standard cube strength of concrete sampled at the mixer, and it will not be constant from one part of a structure to another. In order to assess the strength of a structure, the position and size of reinforcement must be known. A knowledge of reinforcement detail is also of great assistance in the interpretation of crack pattern.

The velocity of sound in concrete is related to its density and elastic modulus, which are related in turn to compressive strength. Ultrasonic pulse velocity test is done by means of soniscope which provides the degree of hardening of concrete. The soniscope is a device for measuring the time of transit of a mechanical pulse through concrete or other material. The pulse is generated and received by means of electrical-mechanical transducers. The interval of time that elapses between the introduction of the pulse and its reception, is measured in microseconds. From the time and length measurement a "Pulse Velocity" is calculated. Pulse velocity increases with the elastic modulus of concrete (as does compressive strength) and hence may be used as a measure of strength. The relationship of pulse velocity versus strength can be established for a particular concrete in question in this way. Soniscope or ultrasonic pulse velocity test is an indication of deterioration. Decreasing pulse velocity is an indication of lower elastic modulus and hence show the sign of deterioration. It should be noted that weak pulse signals indicate interrupted path between transducers and hence indicate a sign of cracking.

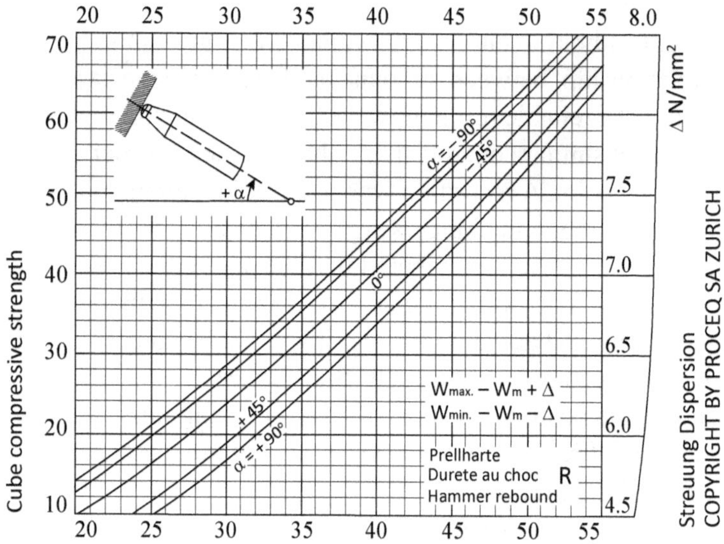

Figure 2.8 Example of Schmidt hammer calibration curves for well-compacted Portland cement concrete with smooth dry surface – note effect when the impact plane is not horizontal (Proceq SA. Zurich)

Care should be taken, also, to choose transmission paths that do not contain reinforcement because the velocity of sound in steel is higher then that in concrete. Non destructive tests have the merit that a large number of readings can be taken fairly quickly and statistical analysis will give an indication of the variability of the concrete and contours of equal value can be plotted. In this way one can locate particularly low results. Experience in using these test results is very helpful and anyone considering them should consult specialized literature on the subject.

In case of ultrasonic pulse velocity test. It is necessary to obtain a reference point by taking readings on a standard specimen such as a cube. In practice, pulses of ultrasound are transmitted through the concrete between two transducers and time taken for the pulses to travel from one to the other is recorded electronically. If the path length is also measured, the pulse velocity can be calculated.

Concrete Pipes

Load calculation for yield and ultimate strength

Length of the pipe = 7.874 foot

Diameter of pipe = 2.986 feet (910 mm)

1500 pounds per linear foot per foot inside dia.

1500	x	7.874	x	2.986	=	35267 pounds
		length of pipe		dia of pipe		(ultimate)

1000 pounds per linear foot dia

1000 x 7.874 x 2.986 = 23516 pounds (yield)

Pipe that support the prescribed load to produce the 0.3 mm crack and do not show a wider crack shall be considered to have met the test requirement.

Three edge bearing test to ultimate load is not required for any class of pipe 1500 mm or less in diameter.

Pipe that have been tested only to the formation of 0.3 mm crack or lesser crack and that meet the 0.3 mm crack load requirement shall be accepted for use.

D. Load: 50 Newton per linear meter per millimeter inside dia (yield)

D. Load: 75 Newton per linear meter per millimeter inside (ultimate)

Concrete class A-2 is mostly used in concrete pipe

Cement	:	sand	:	agg (5 to 20 mm)	according to the concrete mix design established.
465	:	581	:	1046	

Volumetric radio

1 : 1.25 : 2.25

No pipe shall be used within a period of 14 days after curing. All pipes shall be marked with the date of casting.

Steel to be used 58 kg steel of 6mm dia per piece of concrete pipe.

15 + 17 – longitudinal steel bars

23 steel bars spiral (circular)

One concrete pipe 910 mm in dia and 1800 to 1900 kg in weight.

Drainage pipe test: Three point edge bearing test

AASHTO T-280 ASTM C-76, M-170.

Consultant: **RENARDET S.A**

FAISALABAD KHANEWAL MOTORWAY
Faislabad–Gojra Section

Contractor: China Internationl Water and Electoric Corporation

The strength testing of RCC pipe
AASHTO T-280/M-170

Calibration Value=1.0004* Jack Reading

S. No	Nominal Size (mm)	Total Length (m)	Loaded Length (m)	External Diameter (mm)	Internal Diameter (mm)	Wall Thickness (mm)	Proof Load (kg)	Ultimate Load (kg)	Proof Strength N/mm/m	Ultimate Strength N/mm/m
1	910	2.401	2.389	1017	906	103	15006	25010	68.1	113.4

External Load

Crushing strength test by the three edge bearing method the test specimen is tested in a machine designed to apply a crushing force upon the specimen in a plane through the vertical axis extending along the length of the specimen.

Significance and use:

The crushing test can be either a quality control test performed to establish that the finished shippable pipe has sufficient strength to withstand the crushing loads stated in the specifications or a proof design test preformed to prove the adequacy of design. The testing machine may be of any type of sufficient capacity and shall be capable of providing the rate of loading prescribed.

Durability of Concrete

It is essential that concrete should withstand the condition for which it has been designed without deterioration over period of years. Such concrete is said be durable. The absence of durability may be caused either by the environment to which the concrete is exposed or by internal cause within the concrete itself. The external causes can be physical, chemical or mechanical: they may be due to weathering, occurrence of extreme temperature, abrasion, electrolytic action and attack by natural or industrial liquids and gases. The external damage produced by these agents depends largely on the quality of concrete, although under extreme conditions any unprotected concrete will deteriorate.

The internal causes are the alkali – aggregate reaction, volume changes due to differences in thermal properties of aggregate, cement paste and above all the permeability of concrete.

Permeability of concrete

The permeability of concrete is not a simple function of its porosity but depends also on the size, distribution and continuity of the pores. The permeability of concrete is affected also by the properties of the pores. It is also affected by the properties of cement. For the same water cement ratio coarse cement tends to produce a paste with a higher porosity than a finer cement. Permeability determines the relative ease with which concrete can become saturated with water. Permeability of concrete is also of interest in relation of water tightness of liquid retaining and some other structures. Since aggregate particle are enveloped by the cement paste, in fully compacted concrete it is the permeability of the paste that has the greatest effect on the permeability of concrete. The volume of pore space in concrete, as distinct from its permeability is measured by absorption. Absorption no doubt can be considered, a measure of quality of concrete but most good concrete have an absorption well below 10%. The quantity of water flowing through a given thickness of concrete in a given time can be measured and the permeability is expressed as a coefficient of permeability. The permeability test can be performed on cores in order to study the effects of variation in mix proportions and in mixing, placing and curing technique. The test can also give an estimate of the durability of concrete subjected to corrosive action of percolating water.

Air and vapour permeability

Increasing the cement content of the mix has been found to decrease the air permeability, but probably the true factor is the water-cement ratio, it is through a reduction in the actual water-cement ratio that air entrainment reduces the air permeability. The addition of fly ash or pozzolan has a beneficial effect. Grading of aggregate seems to be particularly important in reducing air permeability of concrete.

With very good grading and a cement content of 300 kg/cu.m concrete has been found to be impermeable to air under a pressure of 0.2 Mpa (28 psi), Prolonged curing reduces the air permeability but even at an advanced age rapid drying out increases the permeability, possibly through the development of fine shrinkage cracks.

An in case of air permeability, the water vapour permeability decreases on prolonged moist curing. A decrease in water cement ratio from 0.8 to 0.4 decreases the vapour permeability by 50 to 65 percent.

Chemical attack of concrete

Only a small proportion of concrete used in practice is exposed to serious chemical attack. This is fortunate, since The resistance of concrete to attack by chemical agents is generally lower than to other forms of attack. The most common forms of chemical attack are leaching out of cement and action of sulphates, sea water and natural slightly acidic water. Sea water contain sulphates and in addition to the chemical action crystallization of the salts in the pores of the concrete may result in its disruption owing to the pressure exerted by the salt crystals. Since salt solution rises in the concrete by capillary action, the attack take place only when water can penetrate in to the concrete so that impermeability of the concrete is once again its most important attribute.

In case of reinforced concrete, the absorption of salt establishes anodic and cathodic areas, the resulting electrolytic action leads to an accumulation of the corrosion products on the steal with a consequent rupture of the surrounding concrete, so that the effects of sea water are more severe on reinforced concrete than on plain concrete. It is therefore essential to provide a sufficient cover to reinforcement (50 mm to 75 mm). A cement content of 350 kg/m^3 to 450 kg/m^3 or a water cement ratio of 0.40 to 0.50 are recommended.

2. Effects of frost on fresh concrete.
3. Effects of frost on hardened concrete.

Because of expansion of the mixing water on freezing, a lack of re vibration would allow the concrete to set with a large volume of pore present and consequently the strength of concrete would be very low. If freezing take place after the concrete has set but before it has developed appreciable strength, the expansion associated with the formation of the ice causes disruption and irreparable loss of strength. Control of temperature after placing is obtained by insulating the concrete from the atmosphere and if necessary by constructing enclosures around the concrete and providing a source of heat within the enclosure. The from of heating should be such that the concrete does not dry out rapidly and no part of it is heated excessively.

Frost attack on hardened concrete

As the temperature of saturated hardened concrete is lowered, the water held in capillary pores in the cement paster freezes in a manner similar to the freezing in the capillaries in rock and expansion of the concrete take place. If subsequent thawing is followed by re-freezing further expansion take place so that repeated cycles of freezing and thawing have a cumulative effect and one can envisage analogy between this and fatigue failure. To reduce the danger of frost attack good compaction of concrete is essential and for this reason aggregates and techniques that lead to segregation and honey combing must be avoided. The use of aggregate with a large maximum size or a large proportion of flat particles is inadvisable as pockets of water may collect on the underside of coarse aggregate. There are no established criteria for acceptance or rejection of concrete in terms of the durability factor. The effects of frost can also be assessed from measurements of the loss of compressive or flexural strength or from observation on the change in length.

Effects of De-Icing salts: In case of road slabs frost not only affects the durability of concrete directly but leads also to the use of de-icing salts, which exert an adverse effect on concrete, probably by increasing the severity of the freezing and thawing cycles.

The salts normally used are Nacl and $Cacl_2$ and their repeated applications in intervening periods of freezing or drying results in surface scaling of the concrete. One action of de-icers is to enhance the corrosion of steel. Air entrainment makes concrete more resistant to surface scaling, but the reason for this beneficial influence of entrained air is not clear. Use of rich mixes is also advantageous. Care is required in using Portland pozzolan cement.

Air-entrained concrete

Air entrained concrete offers, however, some advantage not only from the standpoint of durability but also of workability. The use of air-entrained concrete in the future is likely to increase. Entrained air in concrete is defined as air intentionally incorporated by means of a suitable agent. This air should be clearly distinguished from accidentally entrapped air. Entrained air produces discrete cavities in the cement paste so that no channels for the passage of water are formed and the permeability of the concrete is not increased. The air entraining agent can be dispensed either as an admixture, i.e. a material added to the mix when the ingredients are fed into the mixer, or else as an addition to the cement. Entrainment of air has a considerable beneficial effect on the workability of concrete. In order to keep the workability constant, the addition of entrained air can be accompanied by a reduction in the water cement radio, compared with a similar mix without entrained air.

The improvement of workability by the entrained air is probably that air bubbles, kept spherical by the surface tension, acts as a fine aggregate of very low surface friction and considerable elasticity. Entraining air in the mix makes it actually behave like an over-sanded mix and for this reason the addition of entrained air should be accompanied by a reduction in the sand content. The presence of entrained air is also beneficial in reducing bleeding, the air bubbles appear to keep the solid particles in suspension so that sedimentation is reduced and water is not expelled. It is claimed some times that air entertainment reduces segregation. The addition of entrained air lowers the density concrete.

Thermal properties of concrete

Thermal properties of concrete affects its performance over long periods under varying conditions and are also of vital importance in planning mass concrete construction. For this reason data on thermal properties of concrete may be required in design. For instance, a building may require a specific degree of insulation, a slab may have to be free from cracking and warping when subjected to a change in temperature, the basic quantities involved are thermal conductivity, thermal diffusivity, specific heat and co-efficient of thermal expansion.

Resistance of concrete to fire and influence of temperature on strength

Concrete should have good properties with respect to fire resistance. However fire endurance applies really to a building element rather to a building material. In general moisture content of the concrete is the most important factor determining its structural behaviour at high temperature. The loss in strength at high temperature is greater in saturated concrete than in dry concrete and it is the moisture content at the time of application of load that is responsible for the difference.

Resistance to Abrasion and Cavitation

Concrete surfaces are subjected to wear. This may be due to attrition by sliding, scraping or percussion. In case of hydraulic structure, the action of the above materials carried by water generally leads to erosion. Resistance of concrete to abrasion is difficult to assess as the damaging action varies depending on the exact cause of wear.

Erosion resistance: Erosion of concrete is an important type of wear which may occur in concrete in contact with flowing water. It is convenient to distinguish between erosion due to solid particles carried by water and that due to pitting resulting from cavities forming and collapsing in water flowing at high velocities. The rate of erosion depends on the quantity, shape, size and hardness of particles being transported.

In general, at a constant slump, the abrasion resistance increases with a decrease in cement content. In all cases of course, it is only the quality of concrete at and near the surface that is relevant, but even the best concrete will rarely with stand severe abrasion over prolonged periods. An other cause of damage to concrete in flowing water is cavitation. Cavitation damage occurs in open channels generally only at velocities in excess of 12m/s. The surface of concrete affected by cavitation is irregular, Jagged and pitted. Best resistance to cavitation damage is obtained by the use of high strength concrete. The maximum size of aggregate near the surface should not exceed 20mm. Hardness of aggregate is not important but good bond between aggregate and Mortar is vital. Concrete with polymers shows good resistance, depending on the type of the filler, cohesion of binder and adhesion to the filler. Use of steal fibers in the concrete improves resistance to cavitation. The best available guide for selection of abrasion resistant concrete is that the compressive strength of concrete is the paramount factor in its resistant to abrasion. It is usual to specify a maximum cement content of 400kg/m3 with 20mm maximum size of aggregate.

Chemically Resistant Concrete

Silica fume is a by-product of silicon metal production. It has a high content of amorphous silicon dioxide, and consists of very fine spherical particles, on average about 1/100th the size of a particle of Portland cement. When used in combination with super plasticizor, strength increases from 20% to 50%. More importantly, the very fine particles fill up spaces normally available for water to enter the concrete. Without this water, the concrete is less susceptible to freeze– thaw damage. By decreasing the permeability of concrete, it is also difficult for chlorine ions, which would corrode the steel reinforcement, to penetrate the concrete. Another way to prevent corrosion of the steel by chlorides, has been to use stainless and galvanized steel reinforcement bars in critical structures. Recently, these materials have been replaced with a fusion bonded epoxy coating. More economical than stainless steal, these coatings are better suited for salt water than galvanized steel. Though concrete's image is one of the stability and performance, it is actually a fragile material that demands careful handling and more care is needed while using silica fumes in concrete.

Silica fume is added to concrete up to 15% by weight of cement although the normal proportion is from 7 to 10%. It increases the water demand in concrete mix and the problem can be easily compensated by using high range water reducer. It has been observed that silica fume reduces bleeding and it can be successfully used to produce very high strength, low permeability and chemically resistant concrete.

Concrete class A-2

Water cement ratio	0.44
1. Cement	425 kg
2. Silica fumes	30 kg

3. Sand	650 kg
Coarse aggregate	
$^3/_4$"	725 kg
$^1/_2$"	480
Admixture	according to dose recommended
Mixing water	**187 kg- including absorption factor**

Concrete class D-1 (Girders)

Cement	550 kg
Silica fumes	55 kg
Sand	680 kg
Coarse aggregate	
$^3/_4$"	400 kg
$^1/_2$"	610 kg
Admixture	according to dose recommended in order to get workable mix.
Mixing water	195 kg including absorption factor.

The proper way to achieve impermeability in concrete is by lowering the water content and paste content by use of super-plasticizes. To minimize shrinkage and curling the use of high range water reducers is essential.

Concrete Mix Design Class A-2

Construction of barrier with machine
Water cement ratio = 0.32
Ratio of coarse to fine = 62:38
Sargodha Pull # III Chak # 27

1. Source of aggregate.
2. Weight of cement = 470 kg
3. SSD Bulk sp. gr. of coarse aggregate (25mm) = 2.840
4. SSD Bulk sp. gr. of coarse aggregate (20mm) = 2.815
5. SSD Bulk sp. gr. of fine aggregate = 2.720
6. Absorption of coarse aggregate = 0.48%
7. Absorption of fine aggregate = 0.97%
8. Moisture content for coarse aggregate = 0.4%
9. Moisture content for fine aggregate = 0.6%

One cubic meter yield

1. Water cement ratio 0.32
2. Weight of cement 470 kg
3. Volume of cement $V = \frac{M}{D} = \frac{470}{3150} = 0.149206\ M^3$
4. Weight of water = 470 x 0.32 = 150 kg.
5. Volume of water $V = \frac{M}{D} = \frac{150}{1000} = 0.15000\ M^3$
6. Volume of entrapped air = 1.5% = 0.015000 M^3
7. Volume of paste =

Cement	0.149206
Water	0.150000
Entrapped air	0.015000
	0.314206 M^3

9. Total Volume - one cubic meter
10. Volume of coarse and fine aggregate = 1 – 0.314206 = 0.685794 M^3
11. Sp. gr. of coarse aggregate (25 mm) = 2.840 | 2.814 x 0.60 = 1.704
12. Sp. gr. of coarse aggregate (20 mm) = 2.815 | 2.815 x 0.40 = 1.126

13. Sp. gr. of fine aggregate = 2.720
14. Sp. gr. of coarse aggregate (blended) 1.704 + 1.126 = 2.830
15. Sp. gr. of coarse aggregate according to percentage used = 2.830 x 0.62 = 1.755
16. Sp. gr. of fine aggregate according to % used = 2.720 x 0.38 = 1.034
17. SSD Bulk sp. gr. of coarse and fine aggregate = 1.755 + 1.034 = 2.789
18. Wt of aggregate (coarse + Fine) = V x D = 0.685794 x 2789 = 1913 kg
19. Wt of coarse aggregate = 1913 x 0.62 = 1186 kg
20. Wt of fine aggregate = 1913 x 0.38 = 727 = 727 kg

Absorption

1. Absorption of coarse aggregate = 0.48 x 0.62 = 0.298%
2. Absorption of fine aggregate = 0.97 x 0.38 = 0.369%
3. Total Absorption = 0.667%
4. Absorption of water on total aggregate = $1913 \times \frac{0.667}{100} = 12.76$ kg

Moisture content

1. Moisture content of coarse aggregate = 0.4 x 0.62 = 0.248%
2. Moisture content of fine aggregate = 0.6 x 0.38 = 0.228%
3. Total moisture content = 0.476%
4. Moisture content on total aggregate = 1913 x 0.476/100 = 9.11 kg
5. Total water to be used = 150 +12.76 – 9.11 = 153.65 kg

Hence J.M.F Class A-2 construction of Barrier.

1. Cement 470 kg
2. Coarse aggregate (25mm) = 1186 x 0.60 = 712 kg
   Coarse aggregate (20mm) = 1186 x 0.40 = 474 kg
3. Fine aggregate (Lawrencepur sand) = 727 kg
4. Water = 154 kg

As the construction of barrier is carried out with machine high contents of cement have to be used and less water cement ratio (0.30 to 0.32) has to be adopted in order to avoid collapse and cracks in the barrier. Slump of concrete mix to be used is kept on lowest side, maximum slump 2cm is required to be maintained. Care is always adopted to avoid any tilting in the barrier on either side due to which slump of concrete mix is carefully watched at regular intervals. Finishing of barrier is greatly needed at the time of concrete pouring in order to avoid hair cracking, honey combing and blow holes. Ratio between coarse to fine aggregate should be balanced and needs to be carefully decided. The ratio between coarse to fine aggregate needs to be maintained within the range 37 to 40% (fine) : 63 to 60 (coarse). in order to avoid segregation and honey combing.

It is necessary to avoid high contents of cement above 450 kg/m3 except in girders but this is an exceptional case and such quantity of cement 470 kg/m^3 has been used in order to obtain less water cement ratio around 0.32 and to maintain low slump range max. 2cm. However water cement ratio from 0.32 to 0.36 can be maintained with cement content up to 450kg to 460kg/m^3 as water is adjustable according to weather condition which would vary in summer and winter.

Crack Filling in Concrete Structures

Polymer concrete overlays

There are many deteriorated bridge decks in the world to which resurfacing could add ten (10) years of service life. Polymer concrete, basically a mix of resins with silica sand aggregate, can be installed as overlays in a matter of hours. Polymer concrete material is strong and it can give you 4000 psi compressive strength to 6000 psi compressive strength till 24 hours. In addition to high early strength, polymer overlays can be as thin as 1/8 inch. This means less dead weight is added to a bridge. It also means that dock joints and scuppers don't have to be raised after the new surface is applied polymer overlays are also easy to apply as polymer overlay can be laid with ordinary equipment. Polymer concrete overlay is costly and its technology is rapidly changing. There are pilot projects underway in several states on different types of polymer, with different methods of application and with versing properties.

There are three (3) principal resins used for polymer concrete overlays: epoxy, polyester and methyl methacrylates or MMAS. The last category includes high molecule weight methacrylates and modified MMAS.

While most Highway officials and engineers seem to look with favour on polymer overlays, there are many opinions about which type resin is the best as there are research projects underway.

Chad Sherrell is overseeing Washington Dot demonstration projects, with federal Highway Administration support, involving six different types of polymer overlays. His experience began three years ago with Skid resistance problem. Sherrell worked with several manufacturers and used an epoxy-urethane co polymer overlay and his overly can be as thin as 1/4 in.

According to Transpo's Diniit2, MMAS are the best way to go while they are most costly than polyester overlays, they are also stronger, they are less costly than epoxies and the fastest to cure.

Epoxy edge

Epoxy overlays have been around longer, says Dural's Peter Mendis who thinks this ten-year track record can offset epoxy cost higher than either MMAS or polyester. Mendis says epoxy produces a higher strength overlay than does polyester, and that oil spills on the original surface can adversely affect either MMA or polyester overlays but not epoxies which are commonly used at airports Epoxies may be easier to apply as the other polymer overlays are moisture sensitive and require a perfectly dry application surface on the down side, epoxies are less flexible than other kinds of polymer overlays, Several engineers commented that epoxies can deteriorate with ultraviolet exposure. This opinion was questioned by Mendis, Michael Sprinkle of Virginia says polyester cost is about 70 cents per pound while epoxies and MMAs are from two to four times compared to that cost depending on the formulation.

Concrete Sealers

Water proofing of concrete is one practical way to prevent corrosion of embedded steel reinforcement, because by excluding water (an essential element of deterioration) the process is eliminated. Furthermore, in environments in which chloride based deicers are present, transport of dissolved chlorides by the infiltrating water is avoided and thus the major pathway for the corrosion promoting ions to reach the buried steel is blocked. Water proofing, Therefore plays a very important role in the corrosion protection of all types of reinforcing steel.

There are in general, two types of protective materials intended for use on concrete surfaces: membranes and scalers. Membranes are generally impenetrable coatings or layers which can be very effective in preventing moisture ingress. Sealers on the other hand, supplied as a solution or as a suspension in a solvent, merely reduce water inflow. Nevertheless, in some respects scalers offer significant advantages over membranes.

In contrast to membranes, sealers

- do not hamper visual inspection of the concrete.
- do not make the repair of concrete an involved operation.
- are not expensive

All exposed concrete bridge surfaces, whether new or old, precast or field cast are periodically sealed with various approved types of sealers, Exposed concrete decks are sealed with saline sealers, that penetrate and create a hydrophobic surface. The amount of sealer applied depends on the permeability of the deck concrete, the solid content of the Sealer, and the lab test results for the product used. Life expectancy is based on rate of surface abrasion, which is generally a function of traffic volume prior to sealing, the decks are cleaned by washing and than allowed to dry for several days to open the pores in the concrete so the sealer achieves better penetration.

Of all bridge elements, the deck is the most susceptible to deterioration due to its flatness and exposure to salt, ultraviolet rays and abrasion. Sealers are economical at almost any stage of a bridge deck's life, whereas membranes overlays and cathodic protection are Cost-effective at later times on the deterioration curve. Cathodic protection overlays may be used on decks with more advanced stages of reinforcing steel, corrosion, or on deteriorated decks where 30 or more years additional service life are needed.

Cracks allow salt and water to gain easy access to reinforcing steel. Silane or Siloxane sealers have been reported to water proof shrinkage cracks narrower than 0.025 inch (.063 cm).Wider or actively moving cracks may be repaired in several ways, but where snow plow damage to the repair is a concern, the cracks are generally grooved by routing and then filled with either a fully bonded flexible polymer sealant or a debonded high modulus polymer sealent.

Deterioration of asphalt wearing surfaces can be slowed with preventive maintenance. The procedure involves cleaning and drying the surface and all cracks, then treating it with a concentrated asphalt emulsion that fills the cracks, penetrates in voids in the asphalt surface and acts as a binder for a new layer of fine aggregate to restore the skid resistance of the surface.

Actually there are many reports supporting the use of sealers and a recent survey of 45 states in the US showed 28 states using or experimenting with sealers for bridge maintenance. The main reason for the rather slow acceptance of sealers probably is the lack of a standard test procedure for evaluating sealer performance. There is also a communication gap between bridge engineers and sealer chemists. The same problems have slowed the development of preventive maintenance techniques involving chemistry and polymer technology.

A recent report by the state of new York, Highway maintenance division, quantitatively supports the concept of preventive maintenance. A significantly increased maintenance programme is now in operation and being evaluated by the state. Not only is the preventive maintenance economically feasible, it is required in order to optimize the expenditure of public funds.

- Sealers were found to be more effective on good concrete than on inferior concrete. There seems to be no substitute for good quality concrete.
- In view of the detected dependency on characteristics of the concrete, in selecting a sealer for a given application, the properties of concrete have to be considered.
- Only a fraction (35) percent of all the sealers tested penetrate the concrete to a measurable extent.
- Air entrained concrete has a high porosity and apparently higher permeability, resulting in higher water absorption and chloride penetration of coated and uncoated concrete.
- It should be emphasized that these conclusions are based on tests carried out on a limited number of sealers, for relatively short period in the laboratory.

Concrete sealers
Generic type

1. Acrylics
2. Oligomeric Alkoxysiloxanes
3. Epoxies
4. Blends
5. Urethanes
6. Al koxy silanas
7. Silicates/Siliconates
8. Poly siloxanes/siliconates
9. Styrene Acrylate copolymers
10. Chlorinated rubber
11. Siloxane/Methacrylate
12. Oils
13. Synergistic System

Hauling Time of Concrete Mix

Here is another look at maximum mix time (hauling time) requirement for concrete mix. Quality concrete is now being produced with mix time (hauling time) two to three hours but the industry still is customarily specifying the traditional one to $1^1/_2$ hour mix time requirement.

Specification generally refer to ACI 301: "Specification for structural concrete for building". ACI 304: Guide for Measuring, Mixing, transporting and placing concrete and ASTM C-94 standard specification for ready mixed concrete. The paragraphs in these documents that are used every day dealing with slump, entrained air content, total number of revolution on a ready-mix truck counter and the 1 or $1^1/_2$ hour mix time requirement.

Under this time limit if a concrete project is more than 100km from a concrete supplier, it may be necessary to set up a concrete plant on site, arrange cement bags at site with concrete aggregates. The 1 or $1^1/_2$ hour time limit also can be exceeded if there is heavy traffic or a ready mix truck breaks down in transit or if there are mechanical problems with placing the equipment at site. If concrete is rejected from use in structure because of reasons that concrete is not good or because an arbitrary time limit of $1^1/_2$ hour was written in to the concrete specification and concrete gained more hauling time, what would be the answer?

Were did the 1 or $1^1/_2$ hour time limit come from? The section on delivery time of hauling in ASTM C-94-35 "Standard specification for ready mix concrete" states:

"Concrete shall be delivered to the site of work and discharge from the hauling container shall be completed within a period of $1^1/_2$ hour after the introduction of mixing water to dry materials".

This specification was used tentatively in 1933 and adopted in 1935. The 1938 ASTM C-94 reaffirmed the $1^1/_2$ hour mix time but did note that under specially favourable conditions periods up to 2 to 3 hours may be allowed for concrete to be delivered at the site and discharged.

Kindly note that concrete mix for which $1^1/_2$ hour hauling time was allowed, was the concrete without admixture as the cement industry at that time was in the initial stages and no improvement of any kind regarding the concrete admixture was made. Even at that time research has shown no detrimental effects at 1.5 hour time limit. In subsequent years, the two and three hours limit was removed, leaving only the one or $1^1/_2$ hour limits.

Over the years, the cement industry has improved its product. Today's admixture market offers water-reducers, set retarders, super plasticizes and air-entraining concrete admixtures. But the delivery time requirement has not changed in 56 years.

In the remote areas of north-eastern Arizona and northwestern New Mexico, delivery time for concrete – from the time the water and cement come into contact until discharge into the form – often are two to four hours and in some cases even somewhat more due to the distances travelled.

The standard specification for roads and bridges on federal Highway projects (FP-85) which is used by federal agencies in the area allows $1^1/_2$ hours mixing with water reducing/retarding admixture.

With FP-85 as a guideline – Strength, slump, water cement ratio, entrained air content and concrete temperature requirements have been met with concrete delivery at $1^1/_2$ to two hour mix times. This concrete generally has been used on Embankment slope of mitered inlets and outlets of culvert pipes on reservation roads.

On the other hand, deliveries made to non specification jobs using set retarding admixtures have been made up to 4 hours after being loaded, concrete temperature did not exceed 32C°. There was no increased demand for water and the strength was equal to that of concrete delivered at $1^1/_2$ to two hours. However a discrepancy exists in that laboratory research has indicated an increase in temperature time, increased water demand and increased initial and final setting times with these longer mix times.

It is the time to include in specification that concrete mixes can be designed for delivery to projects with in $2^1/_2$ to 3 hours after the cement and water are mixed if retarders are used and if the concrete temperature is no higher than 32C° when placed in to forms and this can be verified with certain trials at site and in the laboratory by taking various samples in test cylinders and crushing such cylinders. Such trials reveal that concrete quality is not adversely affected even if the concrete mix achieves the hauling time from $2^1/_2$ to 3 hours by using certain admixtures and compressive strength of such cylinders provides the specified results. Admixture research has shown that the initial set of concrete can be extended from $2^1/_2$ to 5 hours as compared to initial set of cement with different temperatures. High range water reduces can increase the slump at the project site if required without changing the water cement ratio.

Using admixture in the concrete this hauling time of 1 and $1^1/_2$ hours needs to be put away in history books. The hauling time needs to be increased by using admixtures in the concrete mix.

Pavement Construction

Rigid or Flexible Merits and Demerits

If a Highway design engineer is given the opportunity to make a choice between the type of pavements, rigid or flexible in the area to be designed, he would have to keep in view the various parameters for the safety and protection of the road and particularly the long life of the pavement. If the road is being constructed in marshy places and where there are maximum chances for rain water to accumulate and drainage system is poor, than the decision would certainly go in the favour of rigid pavements. However heavy traffic load is also one of the factors to favour the construction of rigid pavements. No doubt rigid pavements are the concrete roads, highly secured, protected and strengthened but they are very costly, approximately two (2) times as much as compared to the flexible pavements depending upon the class of concrete used and cost of other materials involved. Pavement is always designed to serve the traffic needs adequately over a period of years whether it is rigid or flexible but rigid pavements are constructed for more protection and security and particularly for long life without high maintenance cost. No doubt rigid pavements indicate some demerits but its high economical factor would have to be kept in mind while deciding the choice of rigid or flexible pavements.

Traffic growth in the area is also kept in view while determining structural requirements of the pavements. Traffic growth history for comparable facilities, community and regional planning programs together provide a basis for the estimate. However traffic growth differs greatly between urban and rural Highways. Since average growth percentage is based on total vehicle miles travelled, it is advisable to take into account the separate growth percentage for trucks and passenger vehicles.

Nevertheless the cumulative effects of traffic load are very important factors in the structural design of the pavement. It is very difficult to decide in favour of construction of rigid pavements specially in a country like Pakistan whose poor economy would not allow the construction of such concrete roads. Flexible pavements are preferred to be designed in the countries with poor economy though they are not long lasting and need repair from time to time. However experience has shown that in poor countries like Pakistan, no doubt flexible pavements indicate some damage with the passage of time due to internal and external factors but at the same time such flexible roads are highly economical. Such roads are long distance roads. Though in rigid pavements cement and aggregate provide strength, durability and stability to concrete but such materials are expensive and the designer has to ignore the economy factor while designing such roads.

An adequate and economical design for a pavement structure is just as important as a design for any other engineering structure. Flexible pavements indicate sign of failures due to many road ailments which affect asphalt roads from time to time. This is also due to heavy traffic and poor drainage system and the excess use of unspecified materials. Rutting, bleeding and alligator cracks are the road ailments which affect such roads from time to time due to heavy traffic running specially in one particular lane. These channelized depressions and accumulation of bitumen on top surface at scattered locations cause damage to the road. Alligator cracks are also found developed on the top surface of asphalt pavements due to penetration of rain water or ingress of drainage water in aggregate base course or water bound macadam. Selection of materials that are not the most economical and suitable for the asphalt mix design is contrary to the standards of sound engineering. Construction supervision if very strict and vigilant can play an important role to minimize such road ailment but such vigilant supervision is also necessary during the construction of rigid pavements where water cement ratio need to be highly controlled and much effort is needed to control the quality of cement and aggregate.

An asphalt pavement structure consists of courses of asphalt aggregate mixtures plus any non-rigid courses between the asphalt construction and the foundation or subgrade. The subgrade ultimately carries all traffic loads. Therefore the structural function of asphalt pavement is to support a wheel load on the pavement surface and spread and transfer that load to the subgrade without overtaxing either strength of subgrade or the internal strength of the pavement, Subgrade plays an important role in the construction of rigid pavements also. Full depth asphalt pavements are also constructed to provide more protection and strength to the roads. A full depth asphalt pavement is one in which asphalt mixture are employed for all courses above subgrade or improved subgrade. Full depth asphalt pavements are placed directly on subgrade. Full depth asphalt pavements have many advantages and one of the most important is their ability to resist pavement stresses. Such full depth asphalt pavements are costly and are avoided due to economy and other factors.

Rigid pavements are constructed with concrete mix of 375 kg/M^3 to 450 kg/M^3 cement with water cement ratio 0.45 maximum value. The specified compressive strength of such concrete can be up to 280 kg/M2 whereas the flexural strength which is the most import factor to be considered in this regard is 4.5 Mpa. The flexural strength of concrete is carried out according to ASTM designation C-78-94 (standard text method of flexural strength of concrete) by using simple beam with third point loading. The total length of the beam is 20" whereas only 18" length is used for loading.

Modulus of rupture is calculated

$$R = \frac{PL}{bd_2}$$ – (third point method) where

R = Modulus of rupture PSi or MPa

P = Maximum applied load indicated by the testing machine (Lbs or Newton)

L = Total length of the beam inch or mm.

b = Average width (breadth) of specimen inch or mm

d = Average depth of specimen

$$R = \frac{\overset{P}{38 \times 1000 \times 4572 \text{ mm}} \ \text{KN}}{152.4 \times 152.4 \times 152.4} = 4.91 \text{ MPa}$$

Flexural strength

Flexural tests of moist cured specimens are made as soon as practicable after removal from moist stage. Surface drying of specimen results in a reduction in the measured flexural strength.

$$\text{Compressive strength of cube} = \text{Cube} = \frac{\text{Load}}{\text{area}}$$

if load = 944 kn

Area = 152.4 mm x 152.4 mm

$$\frac{944 \times 1000}{152.4 \times 152.4} = 40.6 \text{ newton/mm}^2 \text{ (MPa)}$$

Slump test of such concrete (rigid pavement) plays an important role in placing the concrete at site as it can be useful as an indication of consistency and also of workability. It is very essential to keep in view the proper control of slump at batching plant and at site. Low slump concrete is necessary for the rigid pavements. However for the pavements in excess of 10 inch in thickness preferably 2″ slump in needed.

However in rigid pavements slump can vary from 2cm to 4cm. Admixture may be used in such type of concrete mix to increase workability and to attain the required slump. Water cement ratio is the most important single factor affecting the flexural strength of concrete and the relationship between the strength and water cement ratio for a given set of materials is in general similar to that for compressive strength. Flexural strength is more sensitive than compressive strength to the aggregate characteristics. It would be preferable to use the maximum size of aggregate one (1) inch, however $1^1/_2$" size of aggregate (maximum size) may also be used. In general as the maximum size of aggregate is increased low water cement ratio can be used for suitable workability and therefore greater strengths are obtained. High flexural strength can only be obtained at a given water cement ratio with smaller size of aggregate. The concrete being used in rigid pavements should be durable and chemically inert under the conditions to which it will be exposed. In case admixture is to be used in such concretes, all necessary tests of admixture need to be done. The complete service record of admixture with related brochure should be obtained from the manufacturer. By lowering the temperature of concrete, retardation is provided to the concrete mix and retarders may not be required if the concrete temperature is below 65F° in some circumstances. So keeping in view the economy factor, the rigid pavements go beyond the approach of Highway Designer financial setup and flexible pavements are preferred due to low cost and expenses.

Asphalt pavements are the quietest on which to drive. Their smooth surfaces offer an absence of noise that is an important safety factor. Noisy pavements, particularly the jointed Portland cement concrete type, on lengthy trips can have a pronounced untoward effect upon the driver's comport, concentration and alertness level. Asphalt pavements are free of repetitive and annoying thumps and surface rumble, and for this reason are the most desirable from the safety standpoint. Pavements that will stand up to punishing traffic loading, that will remain safe when regularly maintained in timely and economical fashion. New pavements, both asphalt concrete and Portland cement, generally have adequate surface skid resistance. However, all pavements wear under traffic, and some worn pavements tend to become slippery, especially when wet. No type of pavement is immune from this condition when prevailing weather brings about the slippery circumstances. In fact, properly constructed asphalt pavements tend to retain higher skid resistance than other types. Open graded mixes have proved excellent in nullifying the danger of water on pavement surfaces.

As a petroleum base product, asphalt concrete is an inert material that is not affected by de-icing agents such as salt. This is not true of Portland cement concrete roads which have a long history of surface spalling and gradual deterioration under repeated applications of salt to control surface icing. Because of the flexible characteristics of asphalt concrete, the pavements structure can conform to minor base and sub grade settlements and still maintain a continuous watertight surface. Portland cement concrete, on the other hand, because of its rigidity, is more prone to failure by cracking and complete fracture with even minor differential sub grade settlements. The flexible characteristics of asphalt concrete significantly contribute to the life expectancy of asphalt pavement with minimum maintenance cost. Flexible pavements contain high adaptability to stage construction with successive pavement layers becoming an integral part of the entire pavement structure and resulting in a substantial increase in the load carrying capacity. Unlike Portland cement concrete asphalt pavements are easily recycled. Both the aggregates and binder may be readily incorporated in to a strengthened pavement.

Review of Project Specifications

It is being realized with passage of time that our project specification are becoming inadequate and not producing the specific standard of asphalt pavements due to high increase of traffic volume day by day.

1. Our sub grade material needs to be improved and strengthened specially in urban areas where A-4 soil is frequently being used without taking into consideration of passing No. 200 sieve and its plasticity index (Pi).whereas sub base and aggregate base course materials also need improvement in order to produce stable and durable roads. When the subgrade is formed in embankment the material placed in the upper part of embankment down to a depth of 30cm below sub grade level should meet compaction requirement of 95% and the soil should provide a minimum value of CBR 15% at 95% compaction of the maximum density as determined by AASHTO T-180 method – D. A-1-a, A-1-b, A-2-4, A3 or A-4 material need to be used according to AASHTO M-145 with P1 not more then 6. Care should be taken that A-4 soil with passing No.200 sieve 70% or less than 70% should be used.
2. The sub base material should provide a CBR value of at least 60% at a density corresponding to 98% of the maximum dry density determined according to AASHTO T-180 method-D.
3. The sub base material retained on sieve No.4 should have a percentage of wear by the Los Angeles abrasion test (AASHTO T-96) of not more than 35%. Such sub base material should have a loss of less than 8% when subjected to five (5) cycles of the sodium sulphate soundness test according to AASHTO T-104. The sand equivalent

determined according to AASHTO T-176 should not be less than 40.

4. Regarding the aggregate base course, the coarse aggregate should have a percent of wear by the Los Angeles Abrasion text (AASHTO T-96) of not more than 30%, Such material should have a loss of less than 8% when subject to five (5) cycles of the sodium sulphate soundness test according to AASHTO T-176. The sand equivalent AASHTO T-176 should not be less than 50. The material passing through 19mm sieve should have a CBR value of minimum 100% tested according to AAHTO T-193. The CBR value should be obtained at the maximum dry density determined according to AASHTO T-180 method-D
5. The physical requirements of coarse aggregate for WBM should satisfy the following limit
   i. Los Angeles Abrasion Max 30%
   ii. Flaky Index Max 15%
   iii. The loss when subject of five (5) cycles of sodium sulphate soundness test (AASHTO T-104) should be less than 8%.
   iv. Coarse aggregate for bituminous paving courses should be cleaned, hard, tough durable and sound and should be of uniform quality and free from decomposed stone, shale, clay, lumps and other deleterious substances. Coarse aggregate for each pavement should be from a single source of supply. Mixing of aggregate from different sources should not be allowed. River gravel should not be approved as a source of supply for coarse aggregate.

The percentage of particles with a minimum of one crushed face should be 100% by weight in asphaltic course. Coarse aggregate in asphaltic course should have properties which comply the following values.

Los Angles Abrasion (ASTM C-131 and C-535)

Asphalt base course 30% Max

Asphalt wearing course	25% Max
Soundness loss (ASTM C-88)	
(Sodium sulphate)	8% Max
Water Absorption (ASTM C 128/127)	2% Max

Fine aggregate for bituminous paving course should consist of the material passing No.4 sieve (4,75 mm)

The material should be non plastic.

Soundness loss (ASTM C-88)	
Sodium sulphate	8% Max

The coarse aggregated in asphalt mix should show no detrimental amount of stripping when tested in accordance with ASTM D-1884. The minimum value of non stripped area would be 95%. If stripping occurs, the aggregate should be rejected or with the agreement of the engineer, a method of treatment proposed for the approval of the engineer to change the material from hydrophilic to hydrophobic state or an approved (anti stripping) additive would be proposed for use with the bitumen binder in accordance with the following criteria. The approved additive should be added to the bitumen material in such a proportion as required to obtain satisfactory results in the affinity with bitumen test performed in accordance with ASTM-1664. The approved additive would be used in strict accordance with the technical specification issued by the additive manufacturer.

Bitumen binder for base and wearing course with penetration grade 60-70 unless otherwise specified on 40-50.

Marshall test criteria for bitumen wearing course

Compaction number of blows each end of specimen	75
Stability (kg)	1000min
Stiffness (stability) Kgf/mm	420
Percent air voids in the mix	4 to 6
Percent voids filled with bitumen	65-75
Percent voids in mineral aggregate	according to table 5.3 MS-2 (asphalt institute USA) 6`th edition 1993
Loss in stability at 60c	20% Max.

Prior to final approval, the proposed job mix (wearing course) with a bitumen content at the permissible upper percentage limit determined in JMF should be compacted to refusal (400/600 blows) and the resulting air voids in the mix should not be less than 2%.

Asphalt Base Course

Prior to final approval proposed job mix of base course with a bitumen content at the permissible upper percentage limit determined in JM# should be compacted to refusal (750/900) and the resulting air voids should not be less than 3%.

Marshall test criteria for asphaltic base course

Compaction: number of blows each end of specimen	75
Stability	1000min
Stiffness (stability/flow) Kgf/mm	430min
Percent air voids in mix	4-7
Percent voids filled with bitumen	55-65
Percent voids in mineral aggregate	

(According to table 5.3 MS-2 (asphalt institute-USA) sixth edition 1993)

Loss in stability at 60c	25%
Filler/bitumen ratio	1:1.5

Bitumen contents for bituminous pavement course

Asphalt base course – 3% min by weight of total mix
Asphalt wearing course – 3.5 min by weight of total mix

Bituminous pavers should be self contained power-propelled units, provided with an automatically controlled activated heated screed or strike-off assembly capable of spreading and finishing courses of bituminous pavement material in width applicable to the specified typical section and thickness shown on the drawings and in accordance with these specifications. Pavers should be equipped with receiving hopper having sufficient capacity for a uniform spreading operation. The paver should be equipped with an automatic feed control, properly adjusted to maintain a uniform depth of material ahead of the screed. The screed or strike-off assembly should be capable of producing a finished surface of required evenness and texture without tearing, shoving or gouging the mixture.

The paver should be equipped with automatic screed controls with sensors on both sides of the paver. Half width paving using one paver should be discouraged. It would be in the interest of the work to use full width paver.

Initial rolling should be commenced when the temperature of the mixture is such that the sum of air temperature plus the temperature of the mixture is between 165 to 190c. In no case shall the temperature of the mixture be less than 125C° on completion of initial breakdown rolling and all other compaction operation should be completed before the temperature of the mixture drops below 100c.

The number and weight of the rollers should be sufficient to obtain the required compaction while the mixture is still at the specified temperature and in workable condition.

All the longitudinal and the lateral joints in bituminous pavements unless formed in an approved method by stop ends should be formed by sawing the laid pavement straight, square and vertical. Hand cutting of joints should not be allowed. Immediately prior to laying of bituminous mixture the joint surfaces should be fully covered by a thin layer of cut back bitumen. Rolling of transverse joints should be both transverse and longitudinal to ensure that a water tight joint is formed.

Regarding the specification of grout of bounded steel, the compressive strength of the hardened grout shall not be less than 170 kg/cm^2 after 7 days. The grout shall be a mixture of OPC cement water and admixture.

The compressive strengths of 300 kg/cm^2 and 350 kg/cm^2 after 7 and 28 days are not feasible as given in different other project specification. It would be in the interest of work to reduce the frequency of coring from the pavement in order to determine compaction and thickness as many holes in the pavements adversely affect the pavement quality. This frequency should be 150m per lane per core instead of 100m per lane per core.

At present Marshall method of 4" dia and 2.5" height has been replaced by Marshall method of 6" dia with 3.75" height with total weight of mould 4000gm, 112 blows on each side of mould, stability 2250 kg minimum, flow 12mm minimum (modified Marshall method as per MS-2 Asphalt institute-sixth edition-1993.

WBM vs Aggregate Base Course

1. WBM is an excellent material for base course if it is properly placed at site and specified rolling is done which includes dry rolling, wet rolling and truck rolling. The interlocking of the particles and proper void filling with fine material khaka/screening needs to be fully ensured. The fine material which consists of crushed stone screening should be free from clay lumps, dirt and other objectionable matter, The percentage of khaka/screening should range from 20 to 30% in WBM at site. It should be kept in mind that the thick layer of khaka on WBM surface must be removed and the corners of the stones should be visible in order to have good bond with asphalt base course in presence of prime coat spray. if it is observed that application of broom is dislocating the WBM coarse particles, hand wire brush needs to be used to remove the layer of khaka.
2. It is a matter of great concern that WBM is a gap graded material and it needs to be dealt with great care and devotion in order to achieve maximum advantage of such material and its supremacy over aggregate base course.
3. WBM is hard, strong, stable and durable material as compared with aggregate base course and provides more protection of the pavement against various road ailments. This is the material which provides more life to road and gives more resistance to the higher traffic load including heavy trucks, tractors and trolleys. WBM is preferred over aggregate base course specially in the areas where heavy axle load and excessive moisture movement/flooding of pavements is expected.
4. WBM is even workable manually or with any available machinery like trucks and tractors whereas aggregate base

course needs graders, loaders and dumpers.

5. It would be essential to use WBM base course on heavily and medium trafficked roads due to the huge increase of traffic volume in Pakistan within the shortest period and accurate judgment of traffic volume becomes difficult. Whereas the use of aggregate base course can be implemented on light trafficked roads due to lesser strength, stability and durability as compared with WBM material.
6. It is not correct to assume that there does not exist a method to check WBM base compaction. The compaction of WBM needs to be checked with the method of T-191, T-180, T-224 except that 2″ retained material should be kept in the field density hole.

   The dia of field density plate and cone should be either 8″ or 12″. In case it is not feasible to return +2″ material in the hole, the volume to +2″ material may be determined based on its specific gravity and the same may be deducted from the gross volume of the test pit hole to find out wet density as under.

$$\text{Wet density} = \frac{\text{wt of material excavated} - \text{wt of } +2'' \text{ material}}{\text{Volume of hole} - \text{volume of } +2'' \text{ material}}$$

The remaining procedure would be according to AASHOT, T-191. Subsequently dry density may be determined and compared with the adjusted MDD to find out the compaction %. Maximum dry density (MDD) should be carried out with 3/4″ passing material and the MDD (Proctor) needs to be corrected by the graph/formula provided in AASHOT, T-224. MMD (Proctor) may be calculated with either direct method (by taking the sample from prepared WBM reach and by sieving the sample through 3/4″ sieve) or replacement method. (2″ pass 3/4″ retained would be replaced by 3/4″ pass & 4 retained).

It is also a matter of great concern that we stress upon the testing of WBM layer more compared to accurate procedure and methodology of placement and compaction of WBM layer. There is no doubt testing is the important phenomenon to check the compaction of WBM course, but at the same time its procedure of placement needs to be strictly followed to ensure interlocking of the coarse aggregate and proper voids filling with khaka or screening. Particularly stages of compaction of WBM including dry rolling, wet rolling and truck rolling need to be strictly followed. So the visual checking of the specified procedure and condition of the field density hole should also be taken into consideration in order to maintain the accuracy in the compaction and placement of WBM layer.

The percentage of khaka from field density hole should also be considered an important factor and it plays an important role in the compaction of WBM course. However retained particles on 2″ sieve and 3/4″ should also be noted.

Good Riding Quality of Pavement

What is the riding quality of the pavement and how the specified riding quality is achieved is a burning topic of the day. The only simple answer which comes to the mind is smooth and uniform driving on the asphalt pavement without any disturbance. Riding quality of the pavement depends upon the maintenance of specified levels of the structural layers of the pavement within specified tolerance. Main stress is on achievement of specified levels of WBM and asphalt courses within the required tolerances. The use of straight edge plays an important role in ensuring the levels of asphalt courses within the required tolerance. However the concerned surveyors should check the levels of WBM, asphalt base course and asphalt wearing course with great care and such levels should be confirmed in view of project specification. The finished profile grade level of the road should be maintained up to the specified standard.

It needs a lot of struggle to maintain the specified levels of WBM but utmost efforts should be done to achieve such specified levels as this is the basic and primary parameter of achieving good riding quality. Asphalt base course needs to be strictly checked with straight edge at the close intervals (12.5 to 25m). Longitudinal and transverse joints also need to be constructed as specified and such joints should be compacted and should come up to the allowable tolerance of straight edge. The surface of asphaltic wearing course needs also to be checked during asphalt paving operation at site by using straight edge at the close interval (6.25 to 12.5M). Longitudinal and transverse joints should be constructed and compacted as specified and such joints should be within the specified tolerance of straight edge. It would be in the interest of work to use the standard paver during asphalt paving operation of base course and wearing course. It would be more beneficial to use the full width paver during such paving operation as it plays an important role in achieving good riding quality of the pavement. It is a matter of great concern that the use of standard type of paver is not stressed upon during asphalt paving operation at site. It has been observed that old, tired and single lane paver is often used at site due to which riding quality of pavement is adversely affected. This is the most important point to note that before paving operation begins, the technical supervision staff of consultant including Resident Engineers and Civil Engineers should make sure that all the paving and compacting equipment made available at site is in proper order without any wear and tear. Full width paver reduces the number of longitudinal joints which provide a great help in achieving good riding quality of pavement. The paver should be equipped with good screed control with the provision of sensor. Use of string line is also important which improves good riding quality of pavement. The rolling equipments need to be of standard quality and of specified weights. The tires of the PTR should be without wear and tear and provide specified tire pressure. The most effective tire pressure for the pneumatic tire roller is the highest pressure that can be used without over stressing the mix. This pressure varies for different mixes, but for normal highway work the maximum pressure is in the range of 90 PSI when the tries are hot. This would mean a 70 to 75 PSI pressure when the tires are cold. This is the most important point to note that consultant supervisory staff should not allow asphalt paving operation at site unless the asphalt pavers and the rolling equipments are of specified standard.

It is a matter of great concern that most of the construction works are carried out in urban areas on emergency basis. Regrettably the asphalt paving operation is often carried out after sunset and continues till late at right due to which the quality of works is adversely affected and specified riding quality of pavement remains unachieved. The client's interest is to finish the work within a very short period due to high pressure on them from political authorities. It has been noticed that such works reflect same kind of deficiency or damage sooner or latter. The concerned authority does not listen to the word of emergency. It is therefore important that principle of safety should be adopted and construction works should only be continued till the time their quality remains unaffected. Asphalt paving operation needs to be discontinued before sunset in order to achieve specified riding quality of the pavement. Quality of works should never be sacrificed for achieving the higher quantities. The completion schedule of the projects should be framed in view of the best quality achievement of the construction work. The concerned Resident Engineers and high officials of consultant team should take stand in such cases and the client staff should not be allowed to interfere in construction in order to finish such works before time. The advice of the client in this regard needs to be taken in to consideration and the decisions should be made in the best interest of the work where quality remains unaffected. It is important to note that completions of work on emergency schedule would certainly affect the quality of work and specified riding quality of the pavement would remain unachieved.

Role of materials in achieving the specified riding quality of the pavements is an important factor which should be also kept in view. Use of WBM class B, no doubt plays an important role in increasing the durability and strength of the pavement but at the same time it affects the riding quality of the pavement up to a certain extent depending upon the methodology of placement and the achievement of specified compaction at site. WBM is an excellent material for the base course especially in areas where high traffic volume is observed and pavements are designed to serve traffic needs adequately over a period of years. It is a matter of great concern that the pavement designers are frequently using the WBM class B material in the structural layers of pavements without keeping in view the volume of traffic analysis in the area and in this way riding quality is being sacrificed at the cost of durability and strength of the road.

It would be more beneficial to use WBM class "C" in the areas where medium to low volume of traffic is observed and especially where high standard of riding quality is required. It is very important to note that WBM class "B" is a gap graded material and it needs to be dealt with great care and devotion in order to achieve maximum advantage of such material and its supremacy over aggregate base course under certain circumstances. No doubt WBM class B is hard, stable and durable material and provides more protection to the pavement against various road ailments and gives more life to the road but at the same time it affects the riding quality of the road specially in the areas where the thickness of asphalt course is 12cm or less than 12cm.

It would be advisable to use aggregate base course material in the areas where light traffic volume is observed and especially where high standard riding quality is needed like urban areas. Such aggregate base course should be laid down on the road with paver to achieve an excellent and best riding quality of the pavement. It is also an important point to note that aggregate base course is used in most of the countries of the world without taking into consideration the traffic volume intensity in the area and such roads have been satisfactorily working for years without showing any significant damage.

So the use of aggregate base course class "A" in producing pavements with good riding quality should be taken into consideration keeping in view the traffic volume in the area.

Kindly note that surveyors and site inspectors including civil Engineers have to play a most important role in achieving good riding quality specially in the areas where WBM materials are used. Use of straight edge should be strictly done at the time of asphalt paving operation at site. Strict survey should be carried out at the time of placing WBM at site. In case proper timings are not followed, specified results would remain unachieved.

Stone Matrix Asphalt Technology Its Implementation in Pakistan

The topic under the discussion is using of stone matrix asphalt on Highways and heavy traffic roads. It is a matter of great concern that AASHTO Designation MP-8 has given various advantages of stone matrix asphalt (SMA) and recommended to carryout the implementation of such asphalt technology in view of its numerous benefits. The binder in stone matrix asphalt needs to be decided according to the climatic conditions of the region. The asphalt binder is based on the mean high and low temperature which the pavement experiences during its life service.

The SMA design of AAHTO MP8-05 is based on the volumetric preparation of the SMA in term of air voids (Va), the voids in mineral aggregate and the presence of stone-on-stone contact. The standard specifies minimum quality requirements for asphalt binder, aggregate, mineral filler and stabilizing additives of SMA mixture design. Stone matrix asphalt is a hot mix asphalt (HMA) consisting of two (2) parts, a coarse aggregate skeleton and a rich asphalt binder mortar. The mixture must have an aggregate skeleton with coarse aggregate-on-coarse aggregate contact (generally referred to as stone-on-stone contact.) The coarse aggregate is generally considered to be that fraction of the aggregate retained on 4 sieve. The air voids range from 3 to 5%. VMA in stone matrix asphalt goes up to 17% minimum. Stability of such asphalt is needed 635 kg (1400 psi) minimum whereas 50 blows by Marshall hammer are required for the preparation of asphalt moulds.

Comparison of Marshall (Conventional) and SMA mix

Sieve size	Marshall Gradation ASTM D3515	SMA Gradation AASHTO MP-8	Marshall Gradation ASTM D-3515	SMA Gradation AASHTO MP-8
25 mm	100	100		
19 mm	90-100	90-100	100	100
12.5 mm	–	50-88	90-100	90-100
9.5 mm	56-80	25-60	–	50-80
4.75 mm	35-65	20-28	44-74	20-35
2.36 mm	23-49	16-24	28-58	16-24
0.300 mm	5-19	–	5-21	–
0.075 mm	2-8	8-11	2-10	8-11

Stone matrix asphalt (SMA) gradation AASHTO MP-8 indicates less fines starting from sieve No. 9.5 mm to sieve No.50 and more filler of passing No. 200 sieve. The gradation envelope specified for stone matrix asphalt (SMA) mix in AASHTO MP-8 gives the higher advantages, no doubt, but its practicality in Pakistan needs to be ensured through various trials in different areas of the country after long running of high traffic volume.

AASHTO MP-8 designates the following advantages for stone matrix asphalt (SMA) and in view of such advantages its importance would remain undeniable.

1. As the medium size aggregates are less, the stone-on-stone contact is easily established.
2. Higher voids in mineral aggregate are achieved.
3. Higher bitumen content properly coat the aggregate which in turn improve the durability of mix.
4. Higher VMA solve the problem of flushing of excess asphalt binder and such higher VMA are achieved due to high percentage of coarse aggregate and less fines.

5. The higher mineral filler along with cellulose fibre acts as stabilizer to prevent the asphalt binder from draining out of the mixture.
6. Flexibility of asphalt mix is also enhanced by high asphalt content and open grade aggregate structure.
7. Durability of asphalt mix is improved by high asphalt binder.
8. The higher asphalt binder results in thick film around the aggregate particles which is more resistant to age hardening and thus reduces the problem or ravelling and pavement brittleness.
9. Higher stability values are not useful at the expense of durability as such values attribute to the rigid pavements which become brittle and crack.

In view of the advantages of stone matrix asphalt, work on the new technology needs to be carried out in perspective of the materials available in Pakistan on ongoing projects in order to ensure positive result of the new technology. It is a matter of great concern that the pavements constructed with stone matrix asphalt would not be very smooth and uniform in texture and the riding quality would remain doubtful. However the behaviour of stone matrix asphalt (SMA) with penetration grade binder under heavy traffic needs to be closely watched in order to make sure the desired results are achieved under the hard and unpredictable traffic conditions in our country.

It is essential on our part to adopt asphalt mixes on coarser side and to reduce the percentage of fines in order to increase the stability and durability of our asphalt pavements. The technology of shifting the asphalt mixes on coarser side remained unadopted due to the attitude of authorities who love the polished, smooth and fine textured roads. No doubt the National Highway Authority has already taken positive steps in the direction but it still needs improvement. The roads with extraordinarily high traffic volume need proper attention and it becomes necessary to use the coarser asphalt mixes in such roads in order to increase the load bearing capacity and durability and to protect such roads from unnecessary damage at later stage. Many materials Engineers and specialists in Pakistan favour this stone matrix asphalt technology but they do not look in to the matter in the light of producing good riding quality roads which is also a basic need in our country. As the stone matrix asphalt produces higher stability roads with no remarkable riding quality, such pavements become rigid and indicate cracking and brittleness with the passage of time, so such technology could not gain a considerable favour in our country.

UPVC Pressure Pipes

BS-3505:1968/PS 3051-1991 Unplasticized Poly Vinyl Chloride

UPVC pressure pipes are manufactured in conformity with BS 3505:1968/PS-3051/1991. These pipes are used for potable drinking water distribution. Such pipes are prepared from materials which are not affected by corrosion. The manufacturing materials of the pipes cost less compared with the steel and cast iron. UPVC pipes are therefore more economical and quality wise better from the engineering point of view.

Such pipes serve the following purposes in various fields.

1. Cold potable water distribution systems
2. Agriculture, Horticulture and Greenhouse irrigation system.
3. Tube wells
4. Fire-ring mains
5. Effluent pipe lines for industrial and rural wastes
6. Slurry pipe lines carrying abrasive and corrosive mine or quarry material.
7. Electric cable safety system.
8. In the roads and bridges for drainage/rainy water outlet.
9. Power station plant
10. Chemical plant installation.

Such pipes should not be used for hot water distribution system specially when the temperature of hot water is beyond 55C° due to the fact that the softening point of UPVC pipe is around 70 to 80C° and such pipes tend to break due to impact of high temperature water flow. UPVC pressing pipes are preferred compared to Mild steel (MS), Cast iron (C), Galvanized iron (GI), Reinforced concrete cement (RCC) and Asbestos cement (AC) pipes due to their better performance and long life. Such pipes a show resistance to elements injurious to health i.e. free from corrosion, abrasion, growth of bacteria, algae and fungi. UPVC pipes show the following advantages in daily life.

1. Installation is easy as they weigh less. Installation time is reduced and UPVC pipes are less expensive.
2. UPVC pipes indicate high flow capabilities as they are non-conosive and inert to chemical or electrochemical resistance.
3. UPVC pipes do not support combustion.
4. UPVC pipes are insulated thermally and electrically.
5. UPVC pipes do not affect the taste, smell of drinking water.

Such pipes are internationally manufactured according to British Standard BS-3505:– 1968/PS-3051:1991. The raw material of such pipes is also according to the standard quality.

UPVC pipes provide a most valuable piping system in comparison with bulky cast-iron piping system. Such pipes are also used in the following industrial and engineering system of building.

1. Underground pipe system for sewerage and disposal in buildings.
2. Soil and waste pipe of municipal engineering.
3. Such pipes prevent mud depositing inside pipe.

UPVC pipes indicate the following properties

1. Corrosion Free.
2. Resistant up to 60C°.
3. High chemical resistance.
4. Sound proof.
5. Smooth internal surface.
6. Self Extinguishing.
7. Non Conductive.
8. Light weight material.

Such pipes fall in various class systems consisting of Class B, Class C, Class D and Class E.

	(6 bar)	(9 bar)	(12 bar)	(15 bar)
wall thickness	2.9/3.4mm	3.5/4.1 mm	4.6/5.3 mm	5.7/6.6 mm

Each class consists of various diameter of pipes and indicate inner diameter and outer diameter.

Generally Class D PVC pipes are used in roads/bridges for drainage and rainy water outlet.

PVC Pipes Class D

Nominal size (inch)	Outer diameter
2″	60.2 – 60.5 mm
3″	88.7 – 89.1 mm
4″	114.1 – 114.5 mm
6″	168 – 168.5 mm

Tests to be performed for UPVC pipe

1. Outer diameter/inner diameter
2. Wall thickness.
3. Sp. gravity
4. Hydrostatic pressure

PVC pipes class D- Nominal size 2”

Sr. #	Description	Avg. Observed Value
1.	Outer Diameter	60.19
2.	Inner Diameter	52.44
3.	Specific Gravity	1.45
4.	Impact Test	Satisfactory
5.	Hydrostatic Pressure Test (43.2 bar for one hour)	Satisfactory

PVC pipes class D-Outer Dia 3″

Sr. #	Description	Avg. Observed Value
1.	Outer Diameter	89.1 mm
2.	Inner Diameter	79.55 mm
3.	Wall Thickness	5.05 mm

PVC pipes class D- Outer Dia 4″

Sr. #	Description	Avg. Observed Value
1.	Outer Diameter	114.48 mm
2.	Inner Diameter	100.88 mm

RESULTS

PVC pipes class D- Outer Dia 6″

Sr. #	Description	Avg. Observed Value
1.	Outer Diameter	168.10 mm
2.	Inner Diameter	149.21 mm

UPVC-Pressure Pipes for Cold Potable Drinking Water As Per BS-3505-1986/PS-3051:1991

			Class B 200 Ft – Head 87 Ibf/in² (6 Bar)			Class C 300 Ft – Head 130 Ibf/in² (9 Bar)			Class D 400 Ft – Head 173 Ibf/in² (12 Bar)			Class E 500 Ft – Head 217 Ibf/in² (15 Bar)		
Nominal Size Inch	Mean Outside Diameter		Average Value	Individual Value		Average Value	Individual Value		Average Value	Individual Value		Average Value	Individual Value	
	Max	Min	Max	Min	Max	Max	Min	Max	Max	Min	Max	Max	Min	Max
	(mm)	(mm)	(mm)	(mm)	(mm)	(mm)	(mm)	(mm)	(mm)	(mm)	(mm)	(mm)	(mm)	(mm)
3/8	17.0	17.3	–	–	–	–	–	–	–	–	–	1.9	1.5	1.9
1/2	21.2	21.5	–	–	–	–	–	–	–	–	–	2.1	1.7	2.1
3/4	26.6	26.9	–	–	–	–	–	–	–	–	–	2.5	1.9	2.5
1	33.4	33.7	–	–	–	–	–	–	–	–	–	2.7	2.2	2.7
1-1/4	42.1	42.4	–	–	–	–	–	–	2.7	2.2	2.7	3.2	2.7	3.2
1-1/2	48.1	48.4	–	–	–	–	–	–	3.0	2.5	3.0	3.7	3.1	3.7
2	60.2	60.5	–	–	–	3.0	2.5	3.0	3.7	3.1	3.7	4.5	3.9	4.5
2-1/2	75.0	75.3	–	–	–	3.5	3.0	3.5	4.5	3.9	4.5	5.5	4.8	5.5
3	88.7	89.1	3.4	2.9	3.4	4.1	3.5	4.1	5.3	4.6	5.3	6.5	5.7	6.6
4	114.1	114.5	4.0	3.4	4.0	5.2	4.5	5.2	6.8	6.0	6.9	8.3	7.3	8.4
5	140.0	140.4	4.4	3.8	4.4	6.3	5.5	6.4	8.3	7.3	8.4	10.1	9.0	10.4
6	168.0	168.5	5.2	4.5	5.2	7.5	6.6	7.6	9.9	8.8	10.2	12.1	10.8	12.4
7	193.5	194.0	6.0	5.2	6.0	8.7	7.7	8.9	11.4	10.1	11.7	13.9	12.4	14.3
8	218.8	219.4	6.1	5.3	6.1	8.8	7.8	9.0	11.6	10.3	11.9	14.1	12.6	14.5
10	272.6	273.4	7.5	6.6	7.6	10.9	9.7	11.2	14.3	12.8	14.8	17.5	15.7	18.1
12	323.4	324.3	8.8	7.8	9.0	12.9	11.5	13.3	17.0	15.2	17.5	20.8	18.7	21.6
14	355.0	356.0	9.6	8.5	9.8	14.1	12.6	14.5	18.6	16.7	19.2	22.8	20.5	23.6
16	405.9	406.9	10.9	9.7	11.2	16.2	14.5	16.7	21.1	19.0	21.9	26.0	23.4	27.0

	Working Pressure		Testing Pressure		
Class	Bar	Ib/in²	Bar	Ib/in²	Meter Head of Water
B	6	87	9	130	60
C	9	130	14	195	90
D	12	173	18	259	120
E	15	217	23	325	150

Inspection and Petro Graphic Analysis of Concrete Aggregates

ASTM C-295

The coarse aggregate consists of crushed or uncrushed stone, gravel or other approved inert materials with similar characteristics or a combination thereof, having clean, hard strong, and durable uncoated particles free from injurious amount of soft, friable, thin elongated and laminated pieces, alkali, organic or other deleterious mater and conforming to the specification requirements.

The coarse aggregate should be of uniform grading with maximum sizes as required for the various classes of concrete. Coarse aggregate should contain not more than 1.5% by weight of material passing No.200 sieve by washing, not more than 5% of soft fragments. Coarse aggregate should have an abrasion loss of not more than 40% at 500 revolution when tested in accordance with AASHTO T-96. When tested in accordance with AASHTO T-104 for five cycles of sodium sulphate soundness test, the maximum loss should not be more than 12%. The handling and storing of concrete aggregates should be such as to prevent segregation or the inclusion of foreign material. Aggregate should be stored on separate platforms at satisfactory location where mixing of surface material is not possible coarse aggregate be separated in to two or more sizes. Different sizes of aggregate should be stored in cold bins or in separated stock piles sufficiently spaced from each other to prevent the material from intermixing.

Coarse aggregate are stored on the ground where the bottom layer of aggregate should not be disturbed. Coarse aggregate should not contain the brownish particles of shale which are broken during the mixing of concrete and affect the compressive strength of concrete. Alkali silica reactive constituents found in aggregates include opal, cristobalite, tridymite, siliceous and some intermediate volcanic glass, chert, glossy to crypto crystalline acid volcanic rocks, synthetic siliceous glasses, some argillites, phyllites, metamorphic grey wackes, rock containing highly metamorphic quartz such as grey wackes, phyllites, schists, gneisses, gneisses granites, vein quartz, quartzite and sandstone. Such constituents cause alkali silica aggregate reaction and damage the concrete at later stage. Potentially deleterious alkali carbonate reactive rocks are usually calcareous dolomites, dolomitic limestones with clayey insoluble residues. Some dolomites essentially are free of clay and some are very fine grained limestones free of clay and with minor insoluble residue, mostly quartz are also capable of some alkali carbonate reaction.

Petrographic examination may also be used to determine the proportion of cubic, spherical, ellipsoidal, pyramidal, tabular, flat and elongated pieces in an aggregate sample. Flat, elongated and thin chip like particles in aggregate increase the mixing water requirements and decrease concrete strength.

Report of the petrographic examination should summarize the essential data needed to identify the sample as to source and proposed use and include a description giving the essential data on composition and properties of the material. The report should record the last procedure employed and give a description of the nature and features of each important constituent of the sample. Notes should be taken during the examination. Each rock type should be described, the relevant feature may include (1) particle shape, (2) particle surface texture, (3) grain size (4) Internal structure, including observation of pore space, packing of grains cementation of grain.

Color, mineral composition, significant heterogeneities, general physical condition of the rock type in the sample are also the necessary features to be considered.

Petrographic examination should also identify the portion of each coarse aggregate that is composed of weathered or other wise altered particles and the extent of that weathering or alteration, whether it is severe, moderate or slight. It should determine the proportion of each rock type in each condition if the concrete in which the aggregate may be used will be exposed to freezing and thawing in a critically saturated condition. Finally porous and highly weathered or otherwise altered rocks should be discriminated because they will be especially susceptible to damage by freezing and thawing and will cause the aggregate portion of the concrete to fail in freezing and thawing. This will ultimately destroy the concrete because such aggregate con not be protected by adequately air entrained sound mortar.

Petrographic analysis of one sample of coarse aggregate is given below. The sample is of Sargodha in origin and to be used in concrete mix. This is the thin section analysis of polarizing microscope. This is to completely describe the petrography and to identify the deleterious continents.

Petrograhic examination of coarse aggregate

1. Dark grey meta-dolerite 87.5%
2. Greenish Grey meta-delerite 8.7%
3. Rhyolite 1.8%
4. Vein calcite 0.7%
5. Vein quartz 0.9%
6. Weathered metadelerite 0.4%

Sample No. 1 (Sargodha)

Thin section No 1A Calcitised Metadolerite

Texture and structure

The rock is medium grained hypidioblastic and sub-blasto porphyritic.

Mineralogy

1. Plagioclase (30.5%) It occurs as subidioblastic to anhedral crystals. It is saussuritised and calcitised.
2. Chlorite: (37.5%) It occurs as light green flakes and aggregates.

3. Calcite: (12.5%) It occurs as anhedral to subidioblastic grains and patches. It also replaces plagioclase.
4. Quartz 3% It occurs mainly as anhedral grains. It shows only moderate strain extinction.

Magnetite/Ilmenite (6.5%) occur as anhedral grains

Hydomica/Illite. (2.0%) They are secondary minerals and formed due to the alteration of plagioclase.

K-Feldspar (4.0%) It occurs as anhedral to subidioblostic grains. It shows alteration to clay.

Epidote (2.0%) It is secondary mineral. Both Zoisite as well as clinozoisite occur. It is anhedral.

Amphibole. (1.5%) It occurs as subhedral crystals.

Sphene/leucoxene. (0.5%) It occurs as anhedral to subidioblastic crystals.

Sample No.1 (Sargodha)

Thin section No. 1B calcitized Metadolerite

Texture and structure

The rock is medium grained hypidioblastic and sub-blasto porphyritic

Minerology

1. Plagioclase: (25.5%) It occurs as subidioblastic to anhedral crystals. It is saussuritised and calcitised.
2. Chlorite: (30.0%) It occurs as light green flakes and aggregates.
3. Calcite (16.0%) It occurs as anhedral to subidioblastic grains and patches. It also replaces plagioclase.
4. Quatz (4.0%) It occurs mainly as anhedral grains. It shows only moderate strain extinction.
5. Magnetite/Ilmonite (5.0%) occur as anhedral grains.
6. Hydromica/Illite (2.5%) They are secondary minerals and formed due to the alteration of plagioclase.
7. K- Feldspar: (30%) It occur as anhedral to subidioblastic grains. It shows alteration to clay.

8. Epidete (8.5%) It is a secondary mineral. Both Zoisite as well as clinozoisite occur. It is anhedral.
9. Amphibole: (4.0%) It occurs as subhedral crystals.
10. Sphere/Leucoxeno (1.5%) It occurs as an hedral to subidioblastic crystals.

Conclusion: Sample No. 1 (Sargodha) Thin section No.1A. Calcitised Metodolerite

The sample contains 2% hydromica. The sample is therefore non deleterious. It has neither alkali silica reaction (ASR) potential nor alkali carbonate (ACR) reaction potential. It may therefore be safely used as an aggregate with ordinary Portland cement/ High alkali cement.

Conclusion: Sample No.1 Thin section No.1B Calcitised Metadolerite

The sample contains 2.5% hydromica. The sample is therefore non-deleterious. It has neither alkali silica reaction (ASR) potential nor alkali carbonate reaction (ACR) potential. It may therefore be safely used as an aggregate with ordinary Portland cement/High alkali cement.

Final conclusion: The petrographic studies show that the sample is non-deleterious. It has neither alkali silica reaction (ASR) potential or alkali carbonate reaction (ACR) potential.

Therefore the sample may be safely used as aggregate with ordinary Portland cement (OPC)/High alkali cement.

Kindly note that if Rhyolite present in concrete coarse aggregate amounts to more them 3% it causes adverse effects to the concrete mix, as Rhyolite is a potentially deleterious constituent with ASR potential. The product of alkali silica reaction is a gel that absorbs water and increases the volume. Pressure generated by the swelling gel ruptures the aggregate particles and causes to extend in to surrounding concrete. Typically alkali silica reaction results in the formation of map pattern cracking of the concrete. Similarly alkali silica reaction (ASR) also occurs due to the presence of vein calcite and vein quartz and produce adverse effects in the concrete at later stage. Weathered metadolerite also causes weakening to the concrete and cause adverse effects to the compressive strength of concrete at a later stage. However weathering to the rock (metadolerite) at a very small scale (less than 1%) would not cause any adverse effect. Hydromica/Illite minerals should not be present in aggregate in more quantity (more then 4%) as such minerals cause adverse effects to the concrete at later stage. It is a matter of great concern. It becomes difficult to distinguish Rhyolite visually in the concrete stockpile aggregate. Petrographic examination of aggregate can only give such distinction. Therefore sampling of aggregate for petrographic examination becomes very important. Sampling should be done by mixing the concrete aggregate with shovel or grader and samples should be selected at random with great care. Such samples should be handed over to the laboratory under the strict supervision.

Visual Inspection and Petrographic Analysis of Fine Aggregate

The fine aggregate consist of sand, sand stone screening or their approved inert materials with similar characteristics or a combination thereof having clean, hard, strong, durable uncoated grains free from injurious amount of dust, lumps, soft or flaky shale, alkali reactive organic mater, loam or other deleterious substances and shall not contain more than 3% of material passing # 200 sieve by washing nor more than 1% of clay lumps or 1% of shale. The use of beach sand should be prohibited. For exposed work, The fine aggregate should be free from any substance that may discolour the concrete surface. The fine aggregate should be uniformly graded and may give the following gradation according to T11 and T-27 – AASHTO standard.

Sieve Designation	% Passing by weight
3/8″	100
# 4	95-100
# 16	45-85
# 50	10-30
# 100	2-10
# 200	0-3

The fineness modulus of the sand should be within the range of 2.3 to 3.1. The variation in fineness modulus of any sand sample should not be more than 0-20 if compared with the fineness modulus of sand used in concrete mix design. The fineness modulus of Lawrancepur sand is normally observed 2.4. The sand Equivalent of sand should not be less than 75% (AASHTO T-17)

Petrography analysis of Lawrancepur sand
One sample collected from the source

Source	Lawrencepur	Sand to be used in concrete mix
Quartz	44.4%	
Amphibole	8.7%	
Quartz mica Schist	7.7%	
Quartz + Polygrain Quartz	7.1%	
Blotite	6.4%	
Carbonate	6.3%	
Feldspar	4.6%	
Granite	4.5%	
Epidote	2.1%	
Magnetite	1.9%	
Slate/Argillite	1.8%	
Muscovite	1.3%	
Greywacke	0.9%	
Acid to Intermediate Volcanics	0.7%	
Chert	0.6	
Lithic arenite	0.3%	
Garnet	0.3%	
Zircon	0.2%	
Tourmaline	0.2%	

Lawrencepur Sand

The deleterious constituents are 0.6% chert 0.9% greywacke 0.7% acid to intermediate volcanics 0.3% Lithic arerite and 1.8% slate/argillite. Slate/argillite are potentially deleterious constituents and such rocky materials contains ASR potential. Alkali silica reaction may cause formation of map pattern cracking in concrete if present in the sand in excess quantity. However these potentially deleterious constituents are within the safe limit. Such type of sample may therefore be safely used as an aggregate with ordinary Portland cement (OPC)/High alkali cement.

Concrete in Girders and Grouting of Bonded Steel

Concrete should be poured in girders whose compressive strength is 350kg/cm^2 (class D-1). Reinforcement steel should be free from injurious defects and its smooth surface needs to be ensured. Steel which shows injurious defects, rusted during or prior to its installation need to be rejected. Strand or rope confirming to ASTM specification A-416 should also be physically checked and such strands for per-stressed concrete need to be tested through an approved laboratory in order to confirm its suitability, Seven wire stress relieved strands should be supplied in coils of sufficient diameter (0.5″ or 0.6″) to ensure that they lie straight.

Grade	Nominal Dia mm	Measure Dia of strand across crown	Yield strength kg	Breaking strength kg	Elongation	Minimum ultimate strength/yield strength
270	12.7 (0.500) inch	12.85 mm	16400	19200	3.5	236245 / 276580

Nominal steel area strands for 0.5" dia $\frac{0.153\ \text{Inch}^2}{98.71\ \text{mm}^2}$

$\frac{16400 \times 2.204}{0.153} = 236245$ PSi yield strength

$\frac{19200 \times 2.204}{0.153} = 276580$ PSi ultimate strength

Grade	Nominal Dia mm	Measure Dia of strand across crown	Yield strength kg	Breaking strength kg	Elongation	Minimum ultimate strength/yield strength
270	12.7 (0.500) inch	12.85 mm	16700	19200	73.5	236245 / 276580

Nominal steel area of strands for 0.6" dia

$\frac{16700 \times 2.204}{0.153} = 240567$ PSi yield strength

Nominal steel area of strand

$\frac{0.217\ \text{inch}^2}{140\ \text{mm}^2}$ $\frac{19200 \times 2.204}{0.153} = 276580$ PSi ultimate strength

The area of strand for nominal diameter 0.5″ (12.70 mm) is 0.153 inch²/98.7²mm whereas the area of strand for nominal diameter 0.6″ (15.24 mm) is 0.217inch²/140mm² steel strands are mostly 270 grade or 250 grade and the specified ultimate compressive strength is 270000 PSI for 270 grade steel strands. Steel to be used in girders also should be tested according to AASHTO M-31 or ASTM A-615. Arrangements should also be made for chemical analysis of steel.

Placing concrete

Concrete should be controlled, mixed and handled as specified. Concrete should not be poured in frames until placing of reinforcement is property inspected. Conduits, anchorages and pre-stressed steel also need to be checked. The concrete slump should be within the range of 140 to 150mm by using standard type of plasticizer. In case the concrete is to be poured at far off places from the concrete batching plant then proper dose of admixture can be used in order to control slump but in no case water cement ratio should go beyond 0.40. In case of not water concrete water chiller should be used or use of excess quantity of ice should be arranged in order to bring the temperature of water below 15 to 10°C. The concrete should be vibrated internally or externally both. The vibrating process should be done with care in such a manner as to avoid displacement of reinforcement, conduits or wires.

Tensioning of per-stressing reinforcement should be commenced unless it is indicated that the concrete has attained the specified strength of 350 kg/cm². After the concrete has attained the required strength, the pre-stressing reinforcement should be stressed by means of jacks to the required tension and stress transferred to the end anchorage. Stressing should be from both ends. The tensioning process should be conducted that the tension being applied and the elongation of per-stressing elements may be measured at all times.

All devices whether hydraulic jack gauges or otherwise should be calibrated so as to permit the stress in the pre-stressing steel to be computed at all times. Safety measures should be taken to prevent accidents due to possible breaking of the pre stressing steel or the slipping of the grips during the pre-stressing process. All equipments should be thoroughly washed with clean water at least once every three (3) hours during the grouting operation and at the end of use for each day. The internal vibration of concrete should be done with great care and the vibrator should reach the bottom concrete in order to carryout the compaction of such concrete as specified. Honey combing and blow holes in concrete should be avoided, honey combing is the result of voids or group of interconnected voids that penetrate more deeply. This is also due to inadequate compaction or leakage of grout in concrete. To avoid honey combing, it is necessary to improve compaction in concrete mix. It is also necessary to reduce maximum size of aggregate if possible and to prevent leakage of grout in concrete. Blow holes in surface of concrete are the result of air bubbles or bubbles of water which are trapped against the face of form work. They may also occur due to inadequate vibration of concrete. To avoid blow holes vibration in concrete should be improved and it is essential to use absorbent form work.

Finishing of concrete girder should be carried out with great care and in such a way that the minimum masonry work is needed in order to cover up the defective points. The top surface of the girders should be smooth neat and clean without any rough texture.

The curing of girders should be carried out with water or with curing compound. The concrete of girders should be kept continuously wet by the application of water for a minimum period of seven (7) days, after the concrete has been placed. Cotton mats, burlaps, rugs, carpets, earth or sand blankets may be used as a curing medium to retain the moisture. The entire surface of concrete should be kept damp by applying water until the surface of concrete is covered with the curing medium. The moisture from the nozzle should not be applied under pressure directly upon the concrete in a quantity sufficient to cause a flow or wash the surface. At the expiration of the curing period the concrete surface should be cleared of all curing mediums.

Compound method of curing consists of a colorless liquid. The compound should be applied with a pressure spray in such a manner as to cover the entire concrete surface with a uniform film and should be of such character that it will harden within 30 minutes after application. The amount of compound applied should be ample to seal the surface of the concrete thoroughly. Power operated spraying equipment should be equipped with an operational pressure gauge and means of controlling the pressure.

The curing compound should be applied to the concrete after the surface finishing operation immediately after the moisture sheen begins to disappear from the surface but before any drying shrinkage or craze cracks begin to appear. Curing compound may be sampled at source of supply. At the job site curing compound containers should be well sealed and should be labelled with the manufacturer's name. Such compound should be tested in the lab by taking cylinders filled with concrete mix and applying curing compound on the concrete cylinder and then crushing such cylinders in the laboratory in order to determine compressive strength of concrete. The comparison of water cured concrete and compound cured concrete can be done in order to check the performance of both.

Grouting

In the pre-stressed girders, the steel is installed in holes of flexible metal ducts cast in the concrete and bonded to the surrounding concrete by filling the ducts with grout. The grout should be a mixture of Portland cement, water and approved admixture. Admixture may be of Sika Intraplast Z whose quantity is 750 gm/50 kg of cement as approved by the manufacturer. Grout of Fospak may be used with the quantity of 225 gm/50 kg of cement. The exact proportion to be adjusted to form a grout having the proper consistency and under no circumstances the water cement ratio should exceed beyond 0.40. Fluidity of grout should be checked and it must be ensured that choking phenomenon at any stage should not be observed in the tubes or ducts. The compressive strength of the hardened grout should not be less than 170kg/cm^2 after seven (7) days at a temperature of 18°C when making preliminary trials for quality. The grout should be mixed for a minimum of two (2) minutes until a uniform consistency is obtained.

It is a matter of great concern that ofter 3 or 7 days the compressive strengths of 230 or 300kg/cm^2 cannot be achieved and similarly compressive strength of 350 kg/cm^2 after 28 days is not an easy task to achieve as given in some project specification. Also keep this thing in mind that grout is a mixture of Portland cement, water and admixture and if any sandy material is used to increase the strength, its flow ability will be reduced and adversely affected. Unpolished aluminium powder may be added in an amount per sack of cement in accordance with the manufacturer recommendation but that contains no ingredient corrosive to steel. However the effect of commercial plasticisers in aggregate form adversely affect the flowability of grouting fluid. Before approval of admixture specially in aggregate form it should be fully ensued that the fluidity of grout is not being adversely affected and the grouting fluid does not become thick. Alumina powder creates expansion in the grout and cavities are properly filled.

All pre-stressing reinforcement to be bonded should be free of dirt, loose rust grease or other deleterious substances, before grouting, the ducts should be free of water, dirt or any other foreign substance.

The ducts should be blown out with compressed air until no water comes through the duct. The grout should be fluid but proportioned so that free water will not separate out of the mix. Unpolished aluminium powder may be added in an amount per sack of cement that will strengthen the grout. Sufficient pressure should be used in grouting to force the grout completely through the duct but care should be taken that rupturing of the ducts does not occur.

Avoid Failures in Structural Concrete

Road construction works

Damage to concrete structures may be due to the following reasons.

1. Inadequacy of concrete mix design
    a. High water cement ratio
    b. Low cement content/substandard quality cement
    c. High workability (high slump)
       which is obtained by using high quantity of water instead of using admixture.
    d. Improper proportion of coarse to fine aggregate
2. Using poor quality materials
3. Heat of hydration
4. Inadequate vibration
5. Hot weather concrete (temperature of concrete mix and atmospheric temperature)
6. Chemical action or corrosion
7. Poor workmanship
8. Lack of cover
9. Poor standard curing of concrete.
10. Form work of concrete
11. Environmental factors

Water cement ratio plays an important role in increasing or decreasing the strength of concrete. High water cement ratio would certainly reduce the strength of concrete whereas low water cement ratio will increase the strength of concrete. Cement is also one of the main ingredients of concrete increase or decrease in its weight would certainly effect the compressive strength of concrete. Concrete mix design therefore be balanced with accurate quantity of cement while the water cement ratio should also be well controlled. The ratio between coarse to fine aggregate should be well established and balanced. In class A-1 concrete where the quantity of cement is normally maintained at 350 to 360 kg/m^3, the ratio of coarse to fine aggregate should be within the range of 38:62/ 40:60 whereas other classes of concrete (A-2, A-3, D-1) the ratio of coarse to fine aggregate should be 38:62. In class A-1 concrete less fine contents cause segregation in concrete due to the less quantity of cement. In class A-1 concrete where the minimum quantity of cement is around 350kg/m^3, its water cement ratio should be well controlled and it should be 0-50 maximum in order to achieve compressive strength of 210kg/cm^2. Similarly in class A-2 concrete, the minimum quantity of cement should be within the range of 425 kg/m^3 to 450kg/m^3 whereas water cement ratio should remain at 0.45 maximum in order to achieve compressive strength of 280 kg/cm^2. Class D-1 concrete is a rich concrete mix which exhibits 510 to 530 kg/m^3 of cement with water cement ratio 0.40 maximum in order to achieve 350 kg/cm^2 compressive strength. Water cement ratio has an established relationship with cement quantity. In case cement content is increased, water cement ratio is decreased. In the same way workability is also related with water cement ratio and cement content. Kindly note that workability of concrete mix should not be increased by increasing quantity of water as it will reduce the compressive strength of concrete. Required workability or slump can be achieved by using a balanced quantity of admixture proposed by the manufacturer.

Poor standard materials cast adverse effects on the quality of concrete. Concrete aggregate should be hard and strong which may provide specified compressive strength. Aggregate to be used in concrete needs to be tested and passed in all the specified tests of project specification. Such aggregate should not contain weak and loose particles including argilaceous shales, weak sandstones and metamorphic grey wackes. Alkali silica reactive constituents including opal, cristobalite. Siliceous and some intermediate volcanic glass, chert, synthetic siliceous glass, argillites, phyllites, quartz, schist, gneisses, vein quartz, excess quantity of rhyolite should not be present.

Substandard quality of cement plays an important role in decreasing the quality of concrete. High alkali cement should be avoided to use in concrete mix. Low alkali cement are those cements in which the % of alkali contents are less them 0.60%. In low alkali cement special clays like China clay, Bentonite and flint stone are used generally. The limestone used for low alkali cement is generally pure limestone containing minor impurities and in some cases such limestone contain low alkali contents. It is a matter of great concern that alkali if present more then 0.60% cause alkali aggregate reaction and damage concrete structures. Alkali in cement at low level (below 0.60%) may bring the safety to concrete but in no case the alkali figure go beyond 1% in order to avoid any risk that may cause damage at later stage.

It is also a matter of great concern that famous and reputed brand of cement should be used. It is also essential to arrange the physical and chemical testing of cement through an approved and famous laboratory under the supervision of inspection staff. Loss of ignition (Loi) of cement should be properly checked and watched continuously. Testing frequency of cement as given in project specification should be properly followed. The stock of cement in store needs to be inspected on daily basis and the temperature of store should not go beyond 25°C. The cement bags in store should not be piled higher than 2.5 meter. These cement bags should not be under the sun as it would increase the temperature of cement due to which heat of hydration would be adversely affected.

Excessive Heat of hydration raises the temperature of concrete. As the concrete cools it will try to contract. If this contraction is restrained, the concrete will be put into tension and if the strain capacity of concrete is exceeded cracks will form. These thermal contraction cracks are often ascribed to drying shrinkage but they form within a week or so after the concrete is placed. The risk of thermal cracking can be reduced by minimizing restrain to contraction and by delaying cooling until the concrete is strong enough to resist the stress induced.

Form work of concrete should be rigid enough not to settle under the weight of concrete but if it is inadequately supported tension cracks may form in the immature concrete.

Inadequate vibration in concrete may cause blowholes and honey combing. It is a matter of great concern that even with good compaction it may be difficult to avoid blowholes and honey combing completely. Concrete mix design have considerable influence on the formation of blowholes. A very cohesive mix tends to produce more blowholes than a less cohesive one. Honey combing is also a sign of either inadequate compaction or loss of grout through joints in the form work or between form work and previously cast concrete.

Environmental factors also play an important role in damage to concrete. Concrete gradually loses strength with increase in temperature above 32°C, The damage may be greater with aggregate such as quartzite with high co-efficient thermal expansion than with those such as limestones with lower thermal co-efficient. Frost attack on concrete may take the form of spalling of the surface or it may cause random cracking. Damage is made more likely by the use of de-icing salts. All such factors are accidental factors and it is difficult to adopt per-cautionary measure in this regard. However Air entrained concrete is far less susceptible to damage by frost then non air entrained concrete.

The most frequent cause of damage to reinforced concrete structures is corrosion of reinforcement and this is usually the result of carbonation of the concrete or the presence of chlorides-carbonation occurs as a result of penetration of Co_2 from the atmosphere. In the presence of moisture this forms carbonic acid which neutralizes the alkalinity of cement matrix. It is suggested that corrosion is unlikely if the chloride content of concrete is uniformly less than 0.4% by wt of cement and highly probable if it exceeds 1% De-icing salts are more harmful than those that are present in concrete from the start as admixture or contamination of aggregate.

Hot weather concrete is the most sensitive concrete and needs more per-cautionary measures due to intensive heat. It is essential to control the temperature of concrete mix which should not go beyond 32°C. It would be necessary that concrete pouring may start in afternoon at 5 p.m approx. The atmospheric temperature should be checked before starting concrete and that should not go beyond 38°C. To avoid serious damage and cracking in hot weather it would be necessary to protect promptly all exposed surface from drying. The temperature of water should be brought down below 10°C by using ice in reasonable quantity or by using chiller. Temperature of aggregate also needs to be lowered and it can be done by shading the aggregate stock pile, sprinkling or fog spraying of aggregate stockpile. By using all these per-cautionary measures, temperature of concrete mix can be controlled and cracking in concrete can be avoided.

Concrete ailments mostly occur due to the hot weather concrete. Cracks that occur soon after concrete has been placed are mostly shrinkage cracks. If cracks appear in an exposed concrete surface very soon after it has been finished or even in some cases before finishing is complete they are termed as plastic shrinkage cracks. They are caused by rapid drying of the concrete surface while the body of concrete is still plastic. Plastic shrinkage cracks occur mostly in horizontal surfaces during hot weather. Excessive rapid evaporation of water from the surface of concrete is the major cause of plastic shrinkage cracking.

Plastic settlement cracks also appear in concrete after the concrete has been placed. The solid in the mixture begins to settle, displacing the water and entrapped air. The water appears as bleeding water on the surface. The settlement continues until the concrete hardens. Reinforcement, embedded hardware, cluster of large aggregate may obstruct the free settlement of concrete causing it to separate and cracks appear in the plastic concrete. They may sometimes be avoided by changing the concrete mix design and by using air entrained concrete and if they do start to form they can be closed immediately by re compacting the upper part of concrete lift. Kindly note that unsuitable concrete mix design may cause a variety of defects ranging from variation in color to severe honey combing and adverse effects on compressive strength of concrete.

Lack of cover on concrete form-work also causes adverse effects on concrete. Sometime reinforcement becomes displaced while concrete is being placed and compacted, with the result that there is less cover than desired on at least one face of concrete. So it is necessary to provide better support for reinforcement. It is also essential to provide more accurate steel fixing.

Uncured or improperly cured concrete can have adverse effects on the compressive strength of concrete. So curing of concrete should be done according to the specified standard. It has generally been experienced that properly cured concrete gives 20% more strength as compared to the uncured concrete. Curing of concrete can be done with water or with curing compound. However water curing is the best curing method if done properly on regulation basis. Curing with curing compound should only be done on vertical structure where standing of water is not possible.

With the passage of time, it is being realised that Designers play the most important role in the road construction design work. Most of the damages occur in the construction work due to improper attention and negligence of the road designers. When such damages occur at site, the blame mostly goes to the improper control of the materials quality work. TST shoulders and median in the center of the pavement are the proper examples which can be considered in this regard. It is no doubt that TST shoulders are more economical but at the same time they bring early damages due to the continuous running of traffic on such shoulders. Moreover strict quality control is not adopted during the construction of such DST/TST shoulders. Heavy rainy water also cause erosion/denudation of such shoulders. Economy factor at the cost of early damages of construction work is not an encouraging phenomenon specially for big and high cost projects. There are many other examples of such negligence which causes huge loss to construction projects. While deciding the classes of concrete and other parameters related to the work, the Designer should be vigilent and careful. Same precautionary measures need to be adopted for asphalt paving work. Kindly note that supervision staff at site and contractors working staff are followers and they have to work according to the path and guideline provided by the designers. Any collapse and damage to the work at site involve all the supervision staff and designer can not keep himself away from such responsibility.

Grading of Aggregate and Workability of Concrete Mix

The term workability is used with reference to concrete mixture to describe the ease or difficulty which may be encountered in placing the concrete in a particular location. It is the synonymous with placeability. Consistency and workability are dependent on the relative quantities of cement, aggregate and water and on the grading of aggregate. The use of air entrainment affects consistency and workability to a high degree. The principal effect of changing the aggregate grading is to change the amount of cement and water needed to make the mixture workable with the desired water-cement ratio. No absolute measure has yet been devised of consistency or workability. Slump test can be useful as an indication of consistency and with certain mixtures also of workability.

In general as the maximum size of aggregate is increased lower water-cement ratio can be used for suitable workability and therefore greater strengths are abstained for a given cement content. Surface characteristics, the shape and size of the particles and gradation of aggregates are of major concern in the strength of concrete. When the water cement ratio is the same and the mixtures are plastic and workable, considerable changes in grading will affect the strength of concrete only to a small degree.

This is to keep in mind that the free water demand increases as the aggregate size decreases. How ever the most positive method of manipulating the water demand of a particular concrete is to use an admixture. Most of the best admixtures significantly reduce the water content. Thus set of retarders and super plasticizers etc are generally suitable and change the characteristics of water, i.e. reduce the surface tension as well as affecting cement hydration. The amount and density of aggregates controls the density of concrete. Concrete with more water will provide low density and concrete with low water content will give high density. So if you move from low to higher density concrete water demand will start decreasing.

Grading of aggregate is important in reducing air permeability of concrete. With very good grading and cement content of 350 kg per cu.m. Concrete has been formed to be impermeable to air under a pressure of 28 PSi. There is evidence that the more the impermeability of the concrete, the greater its durability and there is less chance of harmful chloride ions getting to steel. In some cases cement serves the purpose of admixture and gives more workability. However, more cement will not retard the setting time of concrete nor will it entrain air for best durability and nor provide the early strength.

Segregation in concrete mix also occurs due to large difference in sizes and density of aggregate particles, dropping concrete from right or over vibration.

In case the proportion between coarse to fine aggregate is not up to standard requirement, segregation in concrete mix will occur specially in concrete where cement content is approximately between 325 kg and 350 kg/cu.m (Class A-1). In such concrete of Class A-1, the ratio between coarse to fine aggregate should be 38:62 or 40 to 60. Whereas in other classes of concrete this ratio can vary from 35/40 (fine:) 65/60 (coarse) in order to avoid segregation phenomenon. Keep in mind that a balanced formula of concrete mix needs to be adopted. The proportionate amount of each ingredient should be carefully decided including adequate cement, water and aggregate and specially keeping in mind the workability for properly mixing, transporting, placing and finishing the concrete. Maximum economy possible with materials available should not be neglected. There is significant relationship between the total quantity of mixing water and the volume of the concrete. Experience has shown that the quantity of water per unit volume of concrete of a given consistency is influenced mainly by the size, particle shape and grading of the aggregate and at the end by the amount of entrained air. The quantity of water is generally decided after carrying out all specified tests of aggregates and after complete visual inspection. Aggregate with more or less favourable particle shape or gradation may be found for which the water requirement would depart somewhat from the average. In any case the proper adjustment will need to be made in subsequent trials to arrive at the combination of materials giving the desired workability within the required water-cement ratio.

Los Angeles Abrasion Test of Coarse Aggregate

This test method covers testing the size of coarse aggregate large than 19mm (3/4″) for resistance to degradation using the Los Angeles testing machine.

This test is a measure of degradation of mineral aggregate of standard gradings resulting from a combination of actions including abrasion or attrition, impact and grinding in a rotating steel drum containing 12 steel spheres. As the drum rotates, a shelf plate picks up the sample and the steel spheres carrying them around until they are dropped to the opposite side of the drum, creating an impact-crushing effect. The contents than roll within the drum with an abrading and grinding action until the shelf plate pick up the sample and the steel spheres, and the cycle is repeated. After the prescribed number of revolutions, the contents are removed from the drum and the aggregate portion is sieved to measure the degradation as percent loss.

Apparatus

1. The Los Angeles Machine shall confirm to the requirements of test method C-131.
2. The operation and maintenance of machine shall be as prescribed in test method C-131.
3. Sieves conforming to specification E 11.
4. Balance: A balance or scale accurate within 0.1% of test load over the range required for the test.

Charge: The charge shall consist of 12 steel spheres averaging approximately 47mm (127/32″) in diameter, each having a mass between 390 and 445 gms and having a total mass of 5000 $\pm$ 25gms.

Gradings of test samples

Passing	Retained	Mass of indicated sizes .g		
3″ (75)	2½″ (63)	2500 ± 50		
2½ (63)	2″ (50)	2500 ± 50		
2 (50)	1½″ (37.5)	5000 ± 50		
1½″ (37.5)	1″ (25)		5000 ± 50	5000 ± 25
1″ (25)	¾″ (19)		5000 ± 25	5000 ± 25

Gradation of water bound macadam base

Sieve Designation	Class B percent passing	Class C percent passing
3″	100	
2½″	90–100	100
2″	25–75	90–100
1½″	0–15	35–75
1″	–	0–15
¾″	0–5	0–5

Place the test sample according to the table given above and charge in the Los Angeles test machine and rotate the machine at 30 to 33 r/m for 1000 revolution. After the prescribed number of revolutions, discharge the material from machine and make a preliminary separation of the sample on a sieve coarser than No. 12. Sieve the fines portion on a 1.70 mm (No. 12).Wash the material coarser than No.12 sieve and oven dry 110 ± 5°C to substantially constent mass and determine the mass to the nearest 1g. If the aggregate is essentially free of adherent coatings and dust, the requirement for washing after test may be waived, but drying before the test is always required.

Calculate the loss (the difference between the original mass and the final mass of the test sample retained on No.12 sieve) as a percentage of the original mass of the test sample.

A Look at the quality control of Bearing Pads and comparison before aging and after aging (Neoprene)

Comparison of quality tests – before aging and after aging (Neoprene)

Method Used: ASTM D 412, D 573, D 2240, D 624, D-395 (II), ASSHTO-M251, D 1149

Measurements & Results:

S. #	Tests Item	Unit	Original Properties	Specs	After Aging 100°C x.70 hrs	Change	Specs	Result achieved
1.	Tensile Strength	Kgf/cm²	209	160 min	109	–48%	–15	Not OK
2.	Elongation at Break	%	502	350 min	344	–31%	–50 (300 min)	OK
3.	Hardness (Shore-A)	Pts	60	60 ± 5	65	+05	+10 points	% max OK
4.	Tear Strength	Kgf/cm	53	13 min	–	–		
5.	Compression Set (67°C x 22hrs)	%	13	25 max	–	–		
6.	Long Duration Compression Test (Hydrostatic load Test at 100 tons)	During load, Bulging was observed & bulging pattern seems to be laminate parallelism & no cracks were observed.						
7.	**Ozone Resistance Test:**							
	Test Conditions:							
	Dimension of specimen	2 x 25 x 150 mm						
	No of specimen	03						
	Strain	20%						
	Ozone Concentration	100 ± 10 PPHM						
	Test Temperature	40 ± 2°C						
	Test duration	100 hrs						
	Movement	Static with planetary effect			After completion of test – 100 hours.			
	Observation	Cracks were observed after 8 hrs.			No crack should be observed.			

S. #	Description of tests	Results to Lab. Reg. No. PE/901				Results achieved
		Before Aging		After Aging at 100°C for 70 hours.		
		Results	Specs		Specs	
1.	Tensile Strength (Kg/cm²).	190	160 min	165	–15	–13.2% OK
2.	Elongation at break on 38mm G.L (%)	418	350 min	375	–50 min 300	300 min OK
3.	Hardness Shore type "A"	62–63	60 ± 5	66-67	+10p	72 max OK
4.	Compression Set at 67°C for 22 hours (%)	16.7	25 max	–		
5.	Tear Strength (Kg/Cm).	66	13 min	–		
6.	Low Temperature Brittleness at –40°C for 5 hours	Satisfactory	–	–		

After accelerated aging in accordance with ASTM designation D-573 for 70 hours at 100°C, the elastomer will not show deterioration changes in excess of the following or show the sign of decrease after aging, except hardness which shows sign of increase after aging.

1. Tensile strength –15%
2. Elongation at break –50 (but not less than 300% total – elongation of the material)
3. Hardness, Points +10 (Hardness would show the sign of increase after aging)
4. Shear test (without vertical load) 7 kg/ sq cm. (min)

Sampling of bearing pads should be performed in-accordance with AASHTO M 251-74 as appropriate for the tests required during or immediately after manufacture.

The contractor should furnish to the engineer a certificate by the manufacturers that the elastomer, and fabric (if used) in the elastomeric bearing pads to be furnished conforms to all of the above requirements. The certification should be supported by a certified copy of the results of tests performed by the manufactures upon samples of the elastomer and fabric to be used in the pads. The samples of bearing pads will be selected at random at the point of manufacture or at the point of contractor at the job site. Bearing pads should be available for sampling three (3) weeks in advance of interned use.

The elastomer compound used in the construction of these bearings should contain only virgin crystallization resistant poly chloroprene (neoprene) or virgin natural polyisopre (natural rubber) as raw polymer. All material should be new with no reclaim incorporated in the furnished bearing.

Steel laminates used for reinforcement should be made from rolled mild steel conforming to ASTM A 36/A 36m. ASTM A 570m or equivalent, unless otherwise specified by the purchaser. The laminates should be of the thickness specified by the purchaser or, if left unspecified, should have a minimum nominal thickness of 1.52 mm.

All material tests should be carried out at 23 $\pm$ 2°C unless otherwise noted.

The dimensions of each bearing should be checked, if any dimensions are outside the limits listed in section 6 of this specification, the lot should be rejected.

Testing of Elastomeric Bearing Pads

1.	Tensile Strength	ASTM D-412
2.	Elongation at break %	ASTM D-412
3.	Compression set % 22 hrs at 67°C	ASTM D-395
4.	Tear Strength kg/cm^2	ASTM D-624

5.	Hardness share type 'A'	ASTM D-2240
6.	Ozone resistance 20% strain 100 hrs at 38°C	ASTM D-1149
7.	Brittleness at low temperature	ASTM D-746
8.	**Load Test**	
	Short Duration	AASHTO M-251
	Long Duration	
9.	Bond to metal	ASTM D-429
10.	Resistance to aging	ASTM D-573

Bitumen and Bituminous Mix Durability

Rigidity or stiffness is caused in hot mixes where low penetration asphalt is used. Such low penetration asphalt causes the volume changes making the pavements rigid. Such asphalt mixes give high values of Marshall stability and low flow values. Pavement of such hot mix cracks under heavy volume of traffic.

$$\text{Rigidity} = \frac{\text{Stability}}{\text{Flow}}$$

Cracking apparently originating at the bituminous layer surface is a widespread problem in regions of high ambient temperature. Hardening of bitumen is cause largely by oxidation and volatisation of the lighter fraction in high ambient temperature. Loading effects on the pavements also play an important role in making the bitumen harden. A flexible pavement consists of layers of different material in a structural system and such pavement has to bear the effects of traffic load and environment for a long period of designed life. The foundation of subgrade is adequately protected and vehicle operation costs including safety and comfort are kept within tolerable limits. With the passage of time the structural integrity and service ability of the pavement is reduced and pavement starts cracking and showing disintegration of the bituminous material including other distresses called rutting and corrugation under the prevailing environmental conditions.

A penetration test on the recovered asphalt indicate low penetration (20-25) which reflect significant hardening of 60-70 penetration asphalt during plant mixing. The rutting of the pavement can be attributed to material properties of the asphalt concrete surface course and mix design techniques adopted in construction of the surface layer. The viscosity measurement on the recovered asphalt shows an alarming rote of hardening of asphalt binder during the mix preparation. The recovered asphalt indicate low penetration (20-25) which can repeat potential fatigue cracking problems. In case of viscous deformation, Improvement in the mixture's asphalt cement viscosity can be achieved by addling additives or other grades of asphalt. However asphalt mixtures are influenced by varying temperature ranges. Consequently it is important to select an asphalt cement grade that is suitable for prevailing climate. In case of hot mean annual air temperature of 24°C 60-70/40-50 pen bitumen is used. Where the warm mean annual air temperature is observed within the range of 7°C-24°C, 80-100/60-70 pen bitumen is generally used. Cold mean annual air temperature up to 7°C indicate the bitumen grade 80-100/120-150 pen. However recommended asphalt grades for varying temperature ranges will have to be kept in view in order to arrive at the right selection grade of bitumen. Low grade bitumen causes rigidity in the pavement whereas the high grade bitumen cause bleeding and rutting in the pavement.

Cores taken from cracked pavement indicate in many cases that cracking has not propagated from the underside of the layer but has apparently initiated at the surface and progressed downwards. Surface cracking is closely associated with excessive hardening of the bitumen. The extent to which bitumen hardening is likely to occur during mixing and installation may be evaluated either by the thin film over test (TFOT) or the rolling thin film oven test (RTFOT). In both cases measurements are made of certain bitumen properties before and after the test to determine the effect of the specified treatment. Both the TFOT and RFOT are intended only to provide measures of the likely change in bitumen properties during hot mix production and installation.

Mix with abnormally higher values of Marshall stability and low flow values are often less desirable because pavement of such mixes tends to be more rigid or brittle and may crack under heavy volume of traffic. Pavement durability is its resistance to weathering, cracking and abrasion by traffic. Pavement durability under traffic depends on resistance to vehicle abrasive action, aggregate break down and deflection caused by heavy loads.

In the situation where severe belittlement extends well below the layer surface, it is possible that both tensile stresses at the surface and horizontal shear stress within the layer are significant factors in crack development. In a mix designed, produced and installed to specification, it is unlikely that excessive hardening will extend to a significant depth below the surface. However hardening to some depth below the surface is inevitable and in this situation the surface tensile strains may be of primary importance.

Hydrated lime is an efficient and comparatively cheap means of promoting adhesion between aggregate and bitumen. It may also be effective in reducing the susceptibility of bitumen to hardening. The most durable pavement have enough asphalt to fully coat the aggregate, plug air passages between voids after compaction and still leave enough unconnected voids to take up asphalt if traffic continues to compact the pavement. Without enough voids the uncompressible asphalt will be squeezed to the surface. Too little asphalt, however, produces a thin film and too many connected air voids, the pavement then is porous – such a mix is for an open graded friction course. To prevent rapid oxidation this type of mix is designated to have a thick asphalt film on the aggregate.

Some signs of poor pavement durability are.

1. Ravelling of the surface.
2. Brittleness and early cracking.
3. Cracking leading to pot holes.
4. Loss of displacement of the asphalt film.

Enough asphalt provides a thick film on the aggregate particles and close passages between voids. The asphalt holds the aggregate particles in their after compaction position. It prevents penetration of air and water in to the pavement and slows oxidation.

The use of sound durable rock in the asphalt mix is necessary to produce durable mix. Some rocks such as shales tend to degrade under constant traffic and weathering condition and ultimately result in break up of pavement. Pug mill mixing also degrades rock and may produce excessive fines. Unsound and absorptive aggregate soaks up some of the asphalt that is supposed to act as a binder in the mix resulting in a dry or lean mix. This has the same effect as a mix containing too little asphalt.

Stability is the resistance to deformation of asphalt paving mixtures under imposed load. Indication of poor pavement stability are:

1. Ruts (channelling)
2. Corrugations
3. Shoving.

Factors that influence stability are:

1. The aggregate surface texture
2. Shape and gradation
3. Quantity of asphalt
4. Mix density

Round and polished aggregate are more likely to have lower stability than angular and rough textured aggregates. For a given type and grading of aggregate the stability usually varies with the amount of asphalt mixed with it.

The densest aggregate grading (maximum density gradation) often make poor paving mixes. Gradation of maximum density may not permit the use of enough asphalt for maximum durability and good workability in placing and compaction. Because of the thinner asphalt films associated with low asphalt content, durability in sacrificed. Gradings that closely approach these straight lines on fig:4 must be adjusted away from them, within acceptable limits to increase the VMA values. This allows enough asphalt to be used to obtain maximum durability without bleeding.

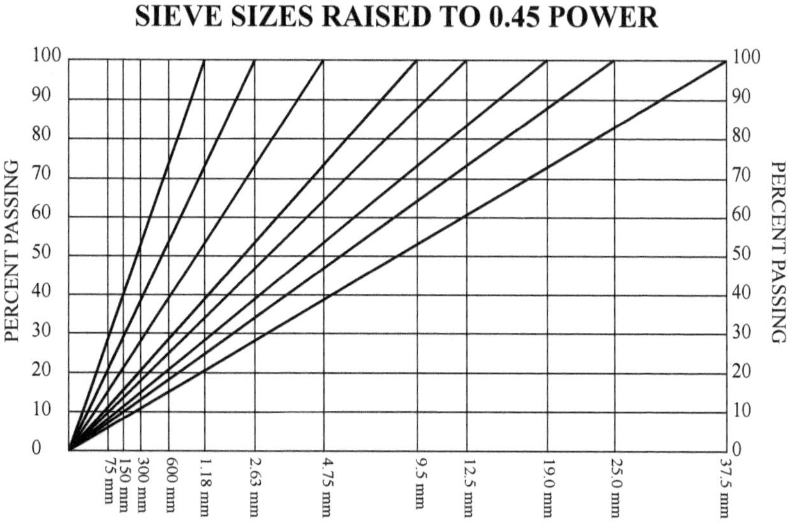

Figure 4: Maximum density curves on Federal Highway Administration 0.45 power gradation chart.

Hydrated lime which is used to reduce the potential for surface cracking in asphalt pavement is effective in reducing aging index of bitumen. Aging index is the ratio of viscosity after aging to initial viscosity.

Malan et al (1989) found the application of coated chippings to be effective in reducing surface cracking. A possible contributory beneficial factor was thought to be the dissipation of residual tensile stresses built in to the surface of the bituminous layer during compaction when rapid cooling occurs and varying resistance to deformation exists throughout the layer depth. It was recommended that pneumatic type rolling be delayed until the mix temperature has dropped below 110°C.

Asphalt Paving Equipment Temperature of Mix

Initial rolling of asphalt mix is started when the temperature of mixture is such that the sum of the air temperature plus the temperature of the mixture is between 165°C to 190°C. In no case the temperature of the mixture should be less than 120°C on completion of initial break down rolling and all other compaction operation should be completed before the temperature of the mixture drops below 90°C. The number and weight of the rollers should be sufficient to obtain the required compaction while the mixture is still at the specified temperature and in a workable condition. The sequence of rolling and selection of roller types should be done by the supervision staff at site to provide the specified asphalt course density. Rollers should move at a low but uniform speed and rolling should be continued until all roller marks are eliminated. Initial or break down rolling should be done by means of either a tandem steel roller which should follow the paver as closely as possible. Rolling should begin as soon as the mixture will bear the roller without undue displacement. The weight of steel wheeled roller (two axle tandem roller) usually ranges in weight from 8 to 14 tons. Intermediate rolling with pneumatic tired roller should be carried out behind the initial rolling. Self-propelled pneumatic tire rollers vary in weight from 12 to 20 tons which is mostly used for asphalt pavement construction. Ballast can be added to the machine to increase the weight. Pneumatic tire rollers have a device to change the tire inflation while the roller is operating. This inflation-on-the run system automatically maintains any preset tire pressure or lower the inflation pressure while the roller is operating. Some compacting conditions and requirements require different inflation pressure. However the temperature which is mostly maintained during intermediate rolling is 90 to 100 psi. Intermediate rolling needs to be started at approximately 130 to 120°C mix temperature. Final rolling (with steel wheel or pneumatic tire roller) should be started at 100 to 110°C and such type of rolling should eliminate the marks of previous rolling.

The temperature of asphalt mix plays an important role in construction of asphalt pavements. Cold weather asphalt paving is most critical and demand more precautionary measures than hot weather asphalt paving where the temperature becomes a controlling factor. The abrupt fall in temperature of the mix from batching plant to site and then the waiting period of such asphalt dumpers till they are laid down at the paving location can have adverse effects on the compaction of the mix. As the initial rolling of the asphalt mix is started late and below the specified temperature, (145 to 140°C) the whole compaction standard including temperature requirements are left unattended and have adverse effects on the compaction of asphalt course. In hot weather temperature of mix is least affected and the downfall of mix temperature from batching plant to site is very little due to which compaction process does not bring much difficulties and problems. However the distance of asphalt batching plant from the site matters more in winter than in summer. More distance involvement would certainly cause more downfall in temperature in winter and asphalt paving process would be adversely affected. So pre-cautionary measures need to be adopted that asphalt dumpers should be laid down at site as soon as they arrive at the site without any wait and delay. Such asphalt dumpers should be released from the batching plant exactly at time when they are needed at site in view of the traveling period. Gathering of asphalt dumpers at site in great numbers and then to put them in the lines due to the delay in site preparation is not a wise step from any point of view. Asphalt paving also demands good management and techniques in order to avoid any complication at later stage. Temperature control is a most important factor in order to produce good asphalt pavements.

Ambient or air temperature has a marked effect on compaction.

The lower the air temperature, the faster the mix cools off. Lift less then 2″ in thickness should not be placed in cold weather at 40°F or less because the mix cools too quickly for densification to be accomplished by rolling. The mix temperature drops faster in thin lifts after the initial or breakdown pass of the roller then before rolling, particularly in cold weather.

It is necessary that at the time of compaction, The mix temperature and air temperature must be high enough to permit the movement of particles during the rolling operation. It is most important that air temperature or ambient temperature should be kept in view in cold weather before starting asphalt paving operation at site particularly during the winter nights at the end of December and January when the ambient temperature is close to the freezing point, asphalt paving operation should be avoided.

The paver should be equipped with a receiving hopper having sufficient capacity for a uniform spreading operation. The paver should be equipped with automatic feed control, properly adjusted to maintain a uniform depth of material ahead of the screed. The screed should be capable of producing a finished surface of the required evenness and texture without tearing, shoving or gonging the mixture. Many screed units have a temper bar which vibrates up and down and strikes off the paving mix so the screed plate can ride smoothly over it. The temper bar also imparts most of the compaction to the mix in the spreading operation. Other screeds vibrate but do not have tamper bars. Vibration may be accomplished with electric vibrators or eccentrically loaded turning shafts. The frequency of vibration may be controlled, Thus helping to obtain a maximum compacting effort. The screed is usually equipped with heaters to prevent the mix from sticking to the screed plate. They are used to heat the screed at the start of paving operations and sometimes during cool, windy days. They should never be used, however, to attempt to heat the mix being delivered to the paver. The screed should be raised and the bottom surface checked for smoothness ad excessive wear. Tampers should be checked for excessive wear, adjustment and proper operation. Excess wear causes a pitted surface in the mat, and improper adjustment gives the mat a scuffed appearance. Automatic screed controls have five (5) main components.

1. Sensor
2. Pendulum
3. Control Box
4. Command Panel
5. Motors or cylinders to change the screed tilt.

The mixing temperature of asphalt should be 160 ± 5 at the batching plant. It necessary that temperature of bitumen at asphalt batching plant should be within the rang of 150°C to 160°C. The temperature of aggregate in dryer should be within the range of 165°C to 170°C, If the temperature of asphalt mix is more than 195°C ± 5°C, It is considered over-heated and be rejected. But if the temperature of asphalt is falling between 163°C to 180°C, it can be used on the condition if the sampled material is properly tested and the overall properties of hot mix are falling within the specified limit of project specification. However precautionary measures should be taken that temperature of asphalt mix may not exceed 160°C in routine work. The temperature fluctuation between the aggregate in dryer and the bitumen in the tank should be around 14°C.

Over burnt mix causes deterioration in the asphalt pavement whereas asphalt mix below temperature of 120°C would not be properly compacted and also cause disintegration.

Silty and Clayey Materials in Highway Construction
Sand Deposits of Burhan/Qibla Bandi/Ghazi Area

Clay generally defined as any very fine grained, natural earthy material which is generally plastic at appropriate water content or hardens when dried or fired.

A-4 and A-5 soils are mainly considered silty soils mixed with clay and minor quantity of fine sand. Such soils are fair to poor whereas A-6 and A-7 soils are clayey soils with poor status. A-4 soils are mostly observed in plain areas of Punjab containing 50% to 70% passing of sieve No.200 but in some cases the passing of sieve # 200 goes up to 80 to 90%. Silty soils fall in the category of A-4 are considered fair to poor from the point of view of Highway construction. However such soils (A-4) provide more strength as compared to the clayey soils (A-6, A-7). Clayey soils (A-6, A-7) indicate failure if any kind of construction is done with such soils including roads and building foundation. A-4 soil which is mostly scattered in plain and deserted areas of Punjab provide better performance in view of Highway construction as compared to clayey soils (A-6, A-7). The maximum dry density of such soils range from 1.95 to 1.80 depending upon the amount of sand mixed, The maximum dry density of clayey soils (A-6, A-7) range from 1.8 to 1.4. A-2 group of soils also contain silty and clayey material mixed with sand and such soils are strengthened with gravel and aggregate particles. The maximum dry density of such soils (A-2-4) goes up to 2.10. The CBR range of A-4 (silty, Clayey with sand mixed) varies from 4 to 20 whereas clayey soils (A-6, A-7) indicate the CBR range from 2 to 5. Clayey soils (A-6, A-7) provide the plasticity index 11 minimum whereas A-4 soil indicate the plasticity index 10 max.

It should be noted that clay is finer than silt characterized by a grain size of less than approximately 4 micrometers. However the term clay can also refer to a rock or a deposit containing a large component of clay size material. Thus clay can be composed of any inorganic material. Clay and silts are considered fair to poor materials but their individual deposits in nature are of great importance according to commercial and Industrial point of view. Various clays possess special properties which make them important industrially. Silty, clayey (A-4) material is also used in Embankment and subgrade.

Silty and clayey soil is also used as sweet soil on the side slopes of Embankment and subgrade for cultivation in order to avoid penetration of rainy and drainage water in the structural elements of pavement. Cultivation of grass, small trees and other crops on the side slopes of Embankment acts as a barrier and allows the water to runaway from the side slopes. Sweet soils are considered costly as compared to Embankment and subgrade material and such soils are also used in agriculture sector for good cultivation purposes. Sweet top soil consists of natural friable surface soil without admixture of undesirable subsoil or foreign material. It is reasonably free from roots, hard clay, coarse gravel, stones larger than 3cm in diameter, noxious wood and tall grass should have been taken from the field where there was a healthy growth of crops, grass, trees or other cultivation.

Clay suitable for the manufacture of heavy clay products, such as building and paving bricks, drain tile and sewer tile are present in several areas of Pakistan including plain and hilly regions and have been utilized for many years. Many deposits of kaolins, ball clay and fire clay are suitable for heavy clay products. To manufacture heavy clay products, the natural material must possess certain working and drying properties. The most important include plasticity when wet, little shrinkage when drying and being fired and good strength. The fired color of the clay is important to both the consumer and manufacture.

The principal clay mineral composing bentonite is smeetite Sodium rich bentonites are valuable for well-drilling muds which must flow when shaken or stirred but harden when left undisturbed. Bentonite occurs as bedded deposits and as a hydrothermal mineral. Bentonite beds that originated as an alteration of water laid volcanic ash, deposited in either fresh or saline water are commercially the most important source of this source. The classification of clay include ball clay, bentonite, bloating clay, brick clay, common clay, fuller earth, pottery clay, fire or refractory clay, high purity kaolins and several others.

Bloating clay may be defined as any fine-grained clayey material which upon firing, expands or bloats in to a frothy mass with closed cellular texture within a limited temperature range. In practice, the definition can be extended to include some shales, slates and mine wastes. After being ground to fine size and mixed with water some of these particles are clayey and can be used as feedstock for making light weight aggregates.

The property of expansion upon firing is desired because blotted material, when used as an aggregate in concrete, reduces the weight of the final product. Two conditions are necessary to bring about bloating.

During firing, when the bloating temperature is reached, the general clayey mass must be in a plastic condition and simultaneously gases must be evolving throughout the mass. Plasticity is developed by the fluxes that start a fusion and, in this state, the gases cannot readily escape. The result is a mass of thin walled cells resembling pumice in appearance. After firing the expounded clay is ground or broken to various sizes and is used in concrete structure, light weight building block, and precast or pre-stressed concrete structural members. The use of light weight aggregates results in considerable monetary saving in structural steel and foundation design on new and existing structures, and in some applications, strength is greater than it would be if heavier rock aggregate were used. ASTM C-330-97. Standard specification for light weight aggregates for structural concrete covers two general types of light weight aggregates. Light weight aggregates are also used for concrete masonary unit – ASTM C-331-98 in which prime consideration is lightness in aggregate.

1. Aggregates prepared by expanding, pelletizing or sintering products such as blast furnace slag, clay, diatomite, flyash, shale or slate.
2. Aggregates prepared by processing natural materials such as pumice, scoria or tuff.

The aggregates shall be composed predominantly of light-weight – cellular and granular inorganic material.

Fuller's Earth or natural bleaching clay is any natural or treated clay which when used as a filter, effectively removes colour and clarifies various minerals and organic oils. The importance of fuller's earth to industry is due to its bleaching properties. Fuller's earth is predominantly smeetite (calcium montmorillonite)

Clays for pottery and stoneware are principally kaolins and/or ball clay. The clay is plastic when wet, shrink little while drying and resist warping and breaking after being dried before firing. After firing, the final product is hard, strong, nonporous and of suitable color. Household items manufactured using pottery clays include porcelain sinks and water closets (toilets), some china ware, various types of over proof cookware, stone ware and other kitchen and dinnerware, ceramic floor, wall tile and decorative items. Ceramic Industries in Gujrat and in other Punjab areas produce many items from such type of clay including kitchen items-dinner sets, tea sets, and many costly floor and wall tiles, Ceramic floor, wall tiles and other decorative items are frequently used in construction of buildings and the clay from which such items are produced is very expensive.

Clay also occurs abundantly in sediments and sedimentary rocks. Clays are a major component of marine sediments. Clays are rarely present as pure minerals, but rather mixtures of different clay types, one group or type normally being dominant. The physical properties of clays, which determine their commercial value, depend on whether or not the various types are admixed, what types of clays are present and their percentage, the % and composition of non clayey materials and other factors. Important physical properties are firing temperature, shrinkage or swelling percentage on firing, plasticity index, blotting characteristics, firing color, percentage of fines (Slack) produced upon crushing, and fired strength.

Utilization of clay as a resource material requires that clay have a restricted range of properties of each industry. Therefore, a brick manufacturer wants a large volume source of moderately uniform clay that has a low shrinkage/swelling percentage, consistent firing colour, relatively low firing temperature and good strength after firing. A paper manufacturer has a specific whiteness test that any given kaoline must pass before it can be used as a surface coating for high quality paper.

Lawrencepur/Qibla Bandi Sand

Sand deposits of Burhan, Qibla bandi, Ghazi area (Attock Dist) have been accumulated by River Indus. Such deposits are of coarser sand/medium coarser sand. Fineness modulus of such sand ranges from 2.7 to 2.9. In many cases such FM comes to be 2.6 to 2.8. However fine sand of fineness modulus ranging from 2.4 to 2.5 is also available. Sand equivalent of such sand ranges from 75 to 80. It is not in the interest of work to use coarsest sand in concrete mix (FM 2-8/2.9) as the finer space of voids are not properly filled due to which honey combing or blow holes arise in concrete structure. It would be in the interest of work to use sand of Fineness modulus ranging from 2.4 to 2.6.

Lawrencepur sand to be used in concrete mix should have the following gradation.

ASTM C 33-97/ T- 11 and T 27

Sieve No	Percent passing	Sieve No	Percent passing
9.5 mm	100	200	0-3
# 4	95-100		
# 8	80-100		
# 16	45-80		
# 30	25-60		
# 50	10-30		
# 100	2-10		

Sand for Back Filling

A-3 material consisting of Ravi sand and chenab sand can be used in back fill where passing #200 sieve does not exceed 10% and such material is also used in back filling of water logged areas as a cushion bed or in such areas where high moisturized soil are encountered. Such high moisture soils are also replaced with sand cushion of A-3 materials. Materials satisfying the requirements of A-3 sand is also used as back fill around structure. Such materials should be non plastic with clay lumps of not more than 1% and also shale of not more than 1%.

Sand has been defined on the basis of gradation leaving aside other properties. However sand will have to qualify all specified tests given in the project specification before use.

The contractor should avoid using aggregate which are alkali silica reactive but in case he wants to use aggregate having alkaline siliceous characteristic, such aggregates would need to be used with blast furnaces slag cement and the work would be without extra-cost and necessary tests of such material would have to be carried out as directed by the project specification or Engineer. Slag cement (GGBFS) acts as admixture, improves workability and finish ability as compared to 100% OPC cement due to increased paste cohesiveness, glassy structure and low water absorption. Slag cement provides higher slump and lowers the permeability of concrete. Slag cement reduces alkali aggregate reaction (ASR) by reacting with alkali in OPC cement and make such alkalis unavailable for reaction. However care should be adopted in using admixture in ground granulated blast cement concrete as blast furnace slag acts as a plasticizer. Laboratory testing and field experience has found slag cement to be compatible with all chemical admixture used in concrete. Slag cement is compatible with pozzolan such as fly ash or silica fumes. Jhelum river concrete aggregate from Nawab shah near Mangla Azad Kashmir contain ASR potential and does not qualify in testing of ASTM C-295 (pectrographic analysis) such material should be avoided. The use of such material in concrete mix can only be permitted if necessary, with low alkali cement/slag cement and cement with additive like stag, fly ash, pozzolan or silica fume.

Asphalt Emulsions

Bitumens are substantially solid at ambient temperature. They must be applied in a fluid state for road construction. This is done by three (3) general methods.

1. By heating the bitumen before its application to aggregate on the road surface. i.e. hot bitumen application.
2. By adding volatile solvent to reduce viscosity. i.e. cut back bitumen application.
3. By the use of asphalt emulsion which dispenses with heating, and cutting back with a solvent and is cheaper than cut back.

Asphalt Emulsion is a two phase system in which asphalt is dispersed in water in the form of tiny globules, There are two types of asphalt emulsion for road construction. They are

1. Anionic Emulsions: Electro negatively charged asphalt globules.
2. Cationic Emulsions: Electro-positively charged asphalt globules depending on the emulsifying agent used.

Cold laid asphalt pavements are generally constructed with mixes made with emulsified asphalt. Aggregate material may be any thing from a dense graded crushed aggregate to a granular soil having a relatively high percentage of dust. The aggregate at the time of mixing may be either damp, air dry or artificially heated and dried. Mixing method may be performed either in the roadway, along the roadway or in a stationary mixing plant. The resulting mixtures usually are spread and compacted at atmospheric temperature however, there are exception. Asphalt mixes made with Mc and Sc-3000 are similar to hot asphalt mixes because they are mixed and placed at temperature in excess of 93°C.

Cold laid asphalt mixes may be used for surface, base or sub base courses if the pavement structure is properly designed. Cold laid asphalt courses are suitable for light and medium traffic.

1. Anionic Emulsions: were developed first for use in road construction. It is an anomic emulsifier and is basic in nature.

The basic disadvantage of anionic emulsion is that after its application, The rate of drying depends on the ambient temperature. Thus the drying time is much more and moreover, in damp condition it cannot be applied.

2. Cationic Emulsion: have been recently developed. It is coming into even more use because of some distinct advantage over anionic emulsion. The rate of drying of cationic Emulsion is chemical irrespective of climatic condition.

Advantages of Asphalt Emulsions:

1. Reduced power fuel requirement: No preheating is required during use.
2. Ease of transport: due to low viscosity and fluidity at ambient temperature, can be directly used for transporting containers.
3. Non-Hazardous: Being made with water.
4. Aggregate: Different types of aggregates can be used. Even damp and wet aggregate are well coated by using asphalt emulsion.
5. Curing time/Dilution: Cationic emulsion cure more rapidly than cut-backs and anomic emulsion. Emulsion can be easily diluted with water.
6. Good Bonding characteristics: Generally asphalt emulsions have better binding characteristic than cut backs.

Cationic Emulsion ASTM D-2397

(1) Rapid setting (2) Medium (3) Slow Setting.

Medium curing Cut-Back asphalt: are produced by cutting back asphalt cement with kerosine which is a moderately volatile product. The presence of kerosine makes the asphalt workable at relatively low temperature, and most of the kerosine in the top layer of a road surface evaporates when exposed to air or heat, leaving the asphalt cement.

Rapid curing cut-back asphalt

Rapid curing cut back asphalt are produced by cutting back asphalt cement with a nephta or gasoline type distillate. Low penetration asphalt cements are usually used for RC than for MC liquid asphalts. This material is rarely used in hot mixing plants because of fire hazard.

Emulsified Asphalts

Ordinarily when it is necessary to mix together two substances such as oil and water, in which one will now remain in suspension in the other for any appreciable length of time, a third ingredient such as soap is added to retard separation. In the same manner asphalt cement and water are mixed using an emulsifying agent to delay separation. Numerous organic and inorganic emulsifying agents are used such as collide clay, soluble or insoluble silicates, soap and sulphonated vegetable oils.

When an emulsion breaks up or sets up, the water drains aways or evaporates leaving the asphalt. Considerable care must be exercised in handling emulsion to prevent premature settling or breaking as excessive pressures, heat or cold. The rapidity of breaking is governed by the proportions and types of constituents in the emulsion. Asphalt emulsions are of two types. The first of these is a standard emulsion in which particles of asphalt are dispersed in water with the aid of emulsifying agent. This is the type most commonly used. An invert emulsion is the second in which very fine globules of water are dispersed in asphalt.

The principal advantages of asphalt emulsion are on cut-back asphalts.

1. Its adaptability if applied to wet aggregates and
2. Elimination of inflammability and toxic hazards.

The type and grade of asphalt selected for use in cold-laid

pavement construction depends a great deal upon the aggregate mixing method and climatic conditions. Cool weather conditions require a lower viscosity grade and permit a faster curing or setting type of asphalt to be selected than for warmer or hot weather mixing conditions. Mixtures prepared with emulsified asphalt do not require aggregates as dry as those prepared with cut back asphalts or road oils. Cold laid asphalt mixtures prepared for maintenance use and stockpiled over a period of time usually are prepared by using one of the lower viscosity grades of the MC or SC cut back asphalts. Aggregates used with emulsions are sometimes put through the dryer to control moisture content if excess water is present. There are four (4) techniques for preparing paving mixtures for cold-laid pavements.

(1) Blade mixing (2) Travel plant mixing (3) Rotary mixing (4) Stationary plant mixing.

TABLE 301-5

REQUIREMENTS FOR EMULSIFIED ASPHALTS (AASHTO M-140)

Type	Rapid Setting				Slow Setting			
Grade	RS-1		RS-2		SS-1		SS-1h	
	Min.	Max.	Min	Max.	Min	Max.	Min.	Max
Test on Emulsions:								
Viscosity, Saybolt Furol at 77°F (25°F). Sec	20	100	-	-	20	100	20	100
Viscosity, Saybolt Furol at 122°F (50°F). Sec	-	-	75	400	-	-	-	-
Settlement 5 days, percent (a)	-	5	-	5	-	5	-	5
Storage stability test 1 day (b)	-	1	-	1	-	1	-	1
Demulsibility, 35 m1. 0.02 $NCaCl_2$ percent	60	-	60	-	-	-	-	-
Cement mixing test, percent	-	-	-	-	-	2.0	-	2.0
Sieve test, percent	-	0.10	-	0.10	-	0.10	-	0.10
Residue by distillation, percent	55	-	63	-	57	-	57	-
Test on Residue from Distillation Test Penetration, 77°F (25°C) 100g. 5 Sec	100	200	100	200	100	200	40	90
Ductility, 77°F (25°C), 5 cm/min.cm	40	-	40	-	40	-	40	-
Solubility in trichloroethylene, percent	97.5	-	97.5	-	97.5	-	97.5	-
Suggested uses	Surface treatment, penetration macadam and tack coat		Surface treatment & penetration macadam		Plant or road mixture with graded and fine aggregates, a substantial quantity of which passes a No.8 (2.3mm) sieve and a portion of which may pass a No. 200 (0.075 mm) sieve, slurry seal teratments.			

a) The test requirement for settlement may be waived when the emulsified asphalt is used in less then 5 days time; or the Engineer may require that the settlement test be run from the time the sample is received until it is used, if the elapsed time is less than 5 days.

b) The 24-hr (1 day) storage stability test may be used instead of the 5 days settlement test.

c) The demulsibility test shall be made within 30 days from date of shipment.

Chemical Requirements – Deformed and Plain Billet Steel Bars for Concrete Reinforcement

ASTM A-165

An analysis of each heat of steel should be made by the manufacture from the test sample taken preferably during the pouring of the heats. The percentage of carbon, manganese, phosphorus and sulphur should be determined. The phosphorus contents thus determined should not exceed 0.05%.

Chemical Composition of Steel bars of various samples

Sample No. 1

S. No.	Elements	Structural Steel	Analysis	Specs%
1.	Carbon	"	0.232	0.26% max
2.	Silicon	"	0.220	
3.	Manganese	"	0.875	
4.	Sulphur	"	0.0310	0.05% max
5.	Phosphorus	"	0.040	0.05% max

Sample No. 2

S. No.	Elements	Structural Steel	Analysis	Specs%
1.	Carbon	"	0.21	0.26
2.	Silicon	"	0.18	
3.	Manganese	"	1.09	
4.	Sulphur	"	0.020	0.05
5.	Phosphorus	"	0.050	0.05

Sample No. 3 Chemical composition

S No.	Elements	Structural Steel	Analysis	Specs%
1.	Carbon	"	0.32	0.26
2.	Silicon	"	0.24	
3.	Manganese	"	0.8	
4.	Sulphur	"	0.034	0.05
5.	Phosphorus	"	0.035	0.05

Sample No. 4 Chemical composition

S No.	Elements	Structural Steel	Analysis	Specs%
1.	Carbon	"	0.37	0.26
2.	Silicon	"	0.30	
3.	Manganese	"	1.09	
4.	Sulphur	"	0.039	0.05
5.	Phosphorus	"	0.042	0.05

The chemical composition of the steel, based on cast analysis shall be in accordance with following table.

Table 1: BS-4449–1997/BS 4461/1978

Element	Grade 250% (max)	Grade 460% max
Carbon	0.25	0.25
Sulphur	0.060	0.050
Phosphorus	0.060	0.050
Nitrogen	0.012	0.012

Maximum carbon equivalent values cast analysis

Table 2: BS 4449–1997/BS 4461/1987

Grade	Maximum carbon equivalent value%
250	0.42
460	0.51

Bars of both grades shall have a carbon equivalent value based on cast analysis, not exceeding the maximum values given in above table.

The following formulas shall be used to calculate the carbon equivalent value, where the chemical symbols represent the percentage of each element.

$$\text{Carbon equivalent value} = C + \frac{Mn}{6} + \frac{Cr + Mo + V}{5} + \frac{Ni + cu}{15}$$

Where

C = % Carbon content

Mn = % Manganese content

Cr = % Chromium content

V = % Vanadium content

Mo = % Molybelenum content

Cu = % Copper content

Ni = % Nickel content

The maximum deviation on product analysis from the values specified for cast analysis on table 1 and table 2 shall be given as under:

Table 3: BS 44449-1997/BS 4461/1978

Element	Deviation above the specified maximum
Carbon	0.02
Sulfur	0.005
Phosphorus	0.005
Total nitrogen	0.001
Carbon Equivalent Value	0.03

Sample for product analysis shall be taken in accordance with the appropriate method of BS 6200.

During product analysis, any bar that falls outside the maximum deviation limits for the composition range of a specified element given in above table shall be deemed not to conform to this British standard.

If during product analysis, a single sample falls outside the maximum deviation limits for the composition range of a specified element given in table 3, further samples shall be selected for analysis from the remainder of the batch.

Water Stops

Water stops should be of polyvinyl chloride (PVC) according to AASHTO-standard specification Division 11 8.9.2.6.2 need to be tested according to AASHTO-standard specification – Division 11 8.9.2.6.4 and should be installed according to specification – Division 11. 8.9.3.4. The frequency of sampling for PVC water-stops would be per lot supplied. The samples of the water stop submitted for testing should be at test one (1) meter long and manufacturer's certificate of the product also needs to be submitted showing that specified specifications are complied with. Methodology of placing water stops and concreting around in vertical and horizontal position should also be followed in view of project specification. Water stops should be of the type, size and shape as shown on the drawings. Water stops should be dense, homogeneous and without holes or other defects. Water stops should be carefully stored on site, away from humidity and sunlight. Back history of water stops supplied by manufactures including the list of international projects where the product was used should also be thoroughly studied. The specified water stops should be used in construction joints, contraction joints and expansion joints of walls, culverts, subway and retaining walls etc. At locations shown on the drawings.

1. Water stops should be free of defects in material handling. It should not be brittle or crack due to normal exposure.
2. Normal water control applications ensure A-hardness should not affect and its characteristics.
3. It should resist normal abrasion and tear failures.
4. It should not fail under normal expansion and contraction in joints if it is installed in a professional way as mentioned in installing method.
5. It should provide effective grip and also effective water barrier.

Secure the water stops in place, using split forms or other suitable method to ensure correct positioning and proper embedment of the water stops in concrete. Position the bulb exactly in the middle of the joint. Vibrate the concrete around the water stop thoroughly to eliminate voids or honeycombs and to ensure effective bonding of the concrete to the water stops ribs. Provide sufficient number of ties to maintain the water stops in position during the second pour.

Water stops

Physical properties and tests of water stops

Property	Test ASTM	Nominal value
Water absorption	*D570	0.02%
Tear resistance	*D624	365 lb./in.
Ultimate Elongation	*D638	350% min
Tensile strength	*D638	2000psi min
Low temperature brittleness	*D746	Passed @-35°F/-37°C
Stiffness in flexure	*D747	700 psi
Specific gravity	*D792	1.40
Hardness Shore A15	*D2240	79 ± 3
Accelerated extraction • Tensile strength • Elongation Effect of alkali • weight change • hardness change	CRD-C 572	CRD-C 572 2100 psi 350% + 0.10% + 1 point

Water stops should be furnished in full length for each straight portion of the joint. Reinforcing bars provided to support the water stops shown on the drawing should be securely held in position by the use of spaces, supporting wires or other approved devices. Water stops should only be used at site when all tests mentioned above come up to the specified standard.

Expansion Joints

Steel for deck expansion joints.
AASHTO – M-160
ASTM – A-6. Rolled steel plates-shapes.
Sheet piling and bars for structural use.

Tension Test
1. Yield strength
2. Tensile strength
3. Bend test.

Chemical analysis will include, a check for the presence of carbon, manganese, phosphorus, sulfur and other minor elements.

The test report shall show heat analysis and results of two tension tests.

The following documents should be submitted before using expansion joints at site:

1. Copy of order placed by the user.
2. Shipment documents.
3. Custom clearance documents.
4. Proof of transportation from port/airport to site.
5. Manufacturer's certificate including test certificate for metal, rubber and bonding between them.
6. Supporting technical data regarding expansion joints.
7. Reference certificate for the last ten (10) years performance.
8. Test certificates from authorized laboratories showing that the proposed joints meet the specification.

The expansion joints should satisfy the following functional requirements.

1. It shall withstand traffic loads of the Highway and accommodate movements between the deck and abutment or the adjacent deck.
2. It shall have good rigidity quality and shall not cause any

inconvenience to road user.

3. It will not be a skidding hazard.
4. It will not generate excessive noise or vibration during the passage of vehicles.
5. Parts liable to wear out shall be easily replaceable.
6. It will be water light and will have provision for carrying away water and silt.
7. It will be easy to inspect and maintain.
8. It will be resistant in hot and very sunny climate.

When directed all materials should be tested before commencement of the bridge work. Test samples need to be tested under supervision of experience consultant supervision staff. Manufactured joints should be delivered in protective and water proof packaging and should be so handled and stored that they remain free from physical damage. The expansion joints should be placed on the basis of manufacturer's recommendation and stated instruction given in the project drawing.

The testing of neoprene (elastomer) is specified as under:

1.	Hardness	ASTM D-2240
2.	Tensile strength at breaking/Elongation at break	ASTM D-412
3.	Deformation under constant compression	ASTM-395
4.	Resistant to aging	ASTM D-573
5.	Resistant to oils (variation in volume)	ASTM D-471
6.	Resistance to ozone	ASTM D-1149
7.	Brittleness at low temperature	ASTM D-746
8.	Bond to metal	ASTM D-429
9.	Tear strength Kg/cm^2	ASTM D-624

Polypropylene Fibre Concrete

This type of concrete is placed before the initial setting time of Rc/Pc concrete for top 2 or 3″ in the construction of Barrage. The ingredients of this concrete is $^1/_2$ inch down coarse aggregate, fine aggregate, cement water and polypropylene fibre. The instruction of the manufacturer of fibre for preparing, placing and curing should be properly followed. Concrete should be treated with propylene synthetic reinforcing fibres as specified by the Engineer in accordance with the fibre manufacturer's recommendation.

1. Synthetic reinforcing fibre should be manufactured from polypropylene to meet ASTM G-116 standard specification for fibre reinforced concrete shortcrete.
2. The dosage is 600gm/50kg bag of cement for concrete works.
3. The contractor is required to make an agreement with fibre supplier for providing technical assistance if required during the work.
4. All fibres should meet the standard and specification according to ACI 544, fibre reinforced concrete. This is the specification mostly used in the concrete construction of barrages.

Polypropylene fibre normally protects the concrete from cracking and this polypropylene fibre is also used sometimes in the concrete of girders D-1 and class A-2 concrete (4000 psi) in minor quantity.

	Fibre tests	Characteristics	Standard/ Designation
	Fibre length	12-25mm	
1.	Compressive strength	5,500-8000	ASTM D-695
2.	Fibre flexural strength	6000-8000	ASTM D-790
3.	Tensile strength at break (PSi)	4500-6000	ASTM D-638
4.	Elongation at break (%)	100-600	ASTM D-638
5.	Water absorption (%)	Negligible (0.01-0.03)	ASTM D-570
6.	Specific gravity	0.90-0.091	ASTM D-792
7.	Ignition point	593°C	
8.	Melting point	160-170	
9.	Heat and UV stabilization	long term	
10.	Thermal conductivity	2.8 10^{-4} Cal cm/	ASTM C-177
11.	Tensile modulus (KSi)	165-225	ASTM D-638
12.	Compressive modulus (KSi)	150-300	ASTM D-695
13.	Flexural modulus @ 25°C	170-250	ASTM D-790
14.	Rockwell hardness	R 80-R 102	ASTM D-785
15.	Electrical conductivity	Low	
16.	Salt resistance	High	
17.	Acid resistance	High	
18.	Alkali resistance	100% (alkali proof)	

Finishing

1. FRC will appear to be slightly stiffer than plain concrete because fibre acts as a cohesive agent and binds the mix together. The workability remains – do not add excess water.
2. Fibres do not float to the surface once they are distributed in the concrete. Fibres may appear on or near surface because they are three dimensionally distributed.
3. Rake should not be used with FRC. Only shovel should be used.
4. For broom or texture finish, broom in any direction.
5. Use of wood, trowels and tool will not be allowed. Only steel or magnesium trowels will be allowed.

Water Proofing

1. Waterproofing consists of asphalt/bitumen or a pre-formed membrane system, both of which include appropriate priming materials.
2. Damp proofing: consists of a coating of primer and two coats of waterproofing asphalt.

Water proofing asphalt conforms to the specification – Asphalt for damp proofing and water proofing ASTM D-312 type I below ground and type II used above ground.

PRIMER: a base material to be used with water proofing asphalt – conform to the specification for primer – Asphalt in damp proofing and water proofing ASTM D41-05.

Fabric: Conform to the specification for bitumen saturated cotton fabric used in roofing and water proofing ASTM D-173-03 or the specification for hot-mixed, hot laid bituminous paving mixtures ASTM designation D 3515-01. The fabric should be stored in a dry and protected place.

Types of water proofing

1. Type I Two coats of hot bitumen over primer.
2. Type II One layer of water proofing membrane over primer.
3. Type III Two layer of water proofing membrane over primer.
4. Type IV Two layers of water proofing membrane over primer with polythene sheets.

Primer: a neoprene based material for use with the rubberized asphalt membrane.

a resin or solvent based material for use with the modified bitumen membrane.

Primer should be of a type recommended by the manufacturer

and approved by the Engineer. Prime surface will not be covered until all solvent constituents have evaporated.

Pre-formed membrane sheet

1. Either the rubberized asphalt type	rubberized asphalt with polyethylene film or mesh.
2. The modified bitumen type	polymer modified bitumen sheet reinforced with a stich-bonded polyester fabric or a fiber glass mesh.

Pre-formed membrane sheet should conform to the following tests requirements:

1. Tensile strength	ASTM D-882
2. % Elongation at break	ASTM D-882
3. Pliability	ASTM D-146
4. Softening point	AASHTO T-53

Mastic: The mastic for use with pre-formed rubberized sheets should be rubberized asphalt cold applied joint sealent. The mastic for use with modified bitumen sheet should be a blend of bituminous and synthetic resins.

Bitumen painting

Two (2) coats of cut back bitumen complying with BS 3690 of viscosity grade within the range of 25 to 50 seconds (Standard tar viscometer) at forty degrees Celsius (40°C) with a coverage of one and eight-tenths (1.8) square meters per liters or any material approved by the Engineer.

All water proofing and damp proofing materials should be tested before use at site. The quality of bitumen to be used should conform to the standard of project specification. The viscosity of cut back bitumen should also be checked in the project laboratory or through an approved laboratory. All materials delivered at site for use should be checked from time to time through surprise visits of supervision staff.

Bitumen Modified Emulsions – Protective Coating

Bitumen emulsified with water, dries to form a black flexible protective coating. The cured film is tough and resistant to the passage of water and vapour transmission. This type of emulsion provides coating for below ground concrete. The emulsion is also used to waterproof walls and floors. It is a cold applied emulsion and is economical in use.

It is easy to apply and provide chemical resistance at all stages. Bitumen contents in the emulsion are up to 60%. Time interval between two coats is approximately 2 hours. The liquid membrane-forming material shall be applied at a rate of 5.0m^2/L (200 ft^2/gal) for testing propose and exposed to air at 23$\pm$2 $^{\circ}$C.50$\pm$10% relative humidity. Application temperature is 5°C. Such emulsions are also available in the market in readymade form from various reputable chemical companies.

ASTM C-309-97 covers liquid membrane forming compounds suitable for application to concrete surfaces to reduce the loss of water during the early-hardening period. White pigmented membrane forming compounds serve the additional purpose of reducing the temperature rise in concrete exposed to radiation from the sun.

All surfaces where the emulsified coating is required for water proofing should be sound and free from contamination such as oil, laitance, grease, wax, dirt or any other form of material which may affect adhesion. One to three coats of bitumen emulsion need to be applied, depending on the degree of protection required. The coverage of bitumen emulsion should be within the range of 1.5 to 4 m^2/lit depending on the substrate. The following types of liquid membrane-forming compounds are included.

Type 1: clear or translucent without dye

Type 1-D: clear or translucent with fugitive dye
Type 2: White pigmented.

The membrane-forming compounds are suitable for use as a curing media for fresh concrete and may also be used for further curing for concrete after removal of forms or after initial moist curing. This is for application to concrete surface to reduce the loss of water during early hardening period.

Improved Subgrade

A Better Choice for Durable and Strengthened Roads

Thickness requirements for asphalt pavements depends to a large extent on the strength of the finished subgrade. The strength of the subgrade mostly depends on the type of material available in such soil and the method to be adopted in order to improve such material. In urban areas the subgrade material mostly consists of A-4 material where passing # 200 sieve varies from 70 to 90%. Whereas in mountaneous areas such subgrade material consists of A-I-a, A-I-b and A-2-4 materials where the improvement of subrade material is not required. The CBR of such materials is mostly more than 30% and in some cases it goes up to 50% depending upon the type of coarser materials available in such soil. However in most urban or desert areas the type of soil available for subgrade is A-4, which consists of silt, clay and a lesser quantity of sand. The CBR of such soil ranges from 8 to 12% at 95% maximum dry density depending upon the sandy and silty material available in such soil. In order to increase the percentage of CBR and bring it to the required standard of improved subgrade, it becomes essential to add sandy material in such soils. The percentage of sandy material to be added in such soils is decided on the basis of certain laboratory trials and CBR tests which indicate the sign of improved subgrade. The subgrade soil is considered to be improved when the percentage of passing # 200 within the limit of 45/50 maximum is achieved; the CBR of such material is achieved within the limit of 25% minimum at 95% maximum dry density. However the concept of testing soils at their anticipated moisture and density should be the governing criteria for thickness design. Stability is closely related to the type of material, its related density and moisture content. It has been commonly observed through various tests of gradation and CBR tests performed that 30 to 50% sandy material (A-3) mixed in the ordinary available subgrade material is required containing 90 to 70% passing of sieve # 200. Such subgrade material can be considered as an improved subgrade and it can provide 25% minimum CBR at 95% maximum dry density with passing # 200 sieve 45% maximum. It is also essential to confirm the classification of sandy material to be added and that should be minimum A-3 material.

The available subgrade material in a given area should be fully analyzed including gradation of material, plasticity index, maximum dry density, grain size hydrometer analysis and CBR test.

After blending of sandy material in common subgrade soil, the whole blended material should be analyzed including its gradation, plasticity index, maximum dry density, hydrometer grain size analysis and CBR test. If common subgrade soil contains less passing of sieve # 200 (50 to 60%) mixing of sandy material would be less in quantity (20% to 30%) in order to achieve specified CBR%.

The blend proportion thus established should apply only when each material to be used is obtained from same source. Should a change in the source of material be made, a new proportion should be established. When unsatisfactory results or other condition make it necessary, additional laboratory tests should be done.

The soil ingredients and water should be mixed in an approved mixing plant (Pug Mill). The plant should be equipped with feeding and metering devices that will add the materials to be blended in the specified quantities. Water should be added during the mixing operation in the quantity required for proper compaction which is approximately optimum moisture content plus or minus 2%. The mixing time should be that which is required to secure a uniform mixture. After mixing the blended material should be transported to the job site while it still contains sufficient moisture and should be placed on the road bed by means of an approved mechanical spreader. The mixture should be spread at rate that will produce a uniform compacted thickness conforming to the required grade and cross-section. Compaction should start as soon as possible after spreading and should continue until the specified relative compaction is achieved.

During the mixing operation, water should be added as necessary to bring the moisture content of the mixture to the percentage suitable for proper compaction.

Road mixed method can also be used. In this case the material should be transported to the site and spread in layers on the road bed in the quantities required to produce the specified blend. The material should be mixed by motor graders and other approved equipment until the mixture is uniform throughout. During mixing operation water should be added as necessary to bring the moisture content to the proper percentage.

Each layer of improved subgrade should be placed in horizontal layers of uniform loose thickness not exceeding twenty (20) centimeters. Each layer should be compacted to the density conforming to the requirements specified.

Improved subgrade is mostly evaluated by CBR test which is probably the most widely used method of designing asphalt pavement structures. The CBR is a comparative measure of the shearing resistance of soil. This test method is intended to provide the relative bearing value (CBR) of base, sub base and subgrade materials. The CBR value for a soil will depend upon its density, molding moisture content and moisture content after soaking.

Improved subgrade provide more strength and durability to the pavement due to better quality of blended material in proper portions. The improved subgrade is not liable to deteriorate or disintegrate due to its good strength and better performance. However care should be taken for better and specified construction.

Bibliography

1. Annual Book of ASTM Standard, Volume 04-02 concrete aggregate C-494, C-295, C-535, C-39-96.
2. Properties of Concrete by A.M. Neville. 3rd edition.
3. Concrete International. Design and Construction.
   i. Extended Mix Time Concrete by Frank. A. Kiziliski.
   ii. Laboratory Investigation of Concrete Sealer by Claire. T. Aitken and Gerard. G. Litvan.
4. Civil Engineering October 1986.
   Filling the Cracks by Virginia-Fairweather.
5. AASHTO Volumes 1 & 2 (M-160, M-170).
6. Asphalt Institute Education series 3 & 7.
7. BS-3505-1968/PS 3051-1991.
8. Asphalt Technology and Construction – Asphalt Institute.
9. Annual Book of ASTM Standard – Part IV Structural Steel. A-615, A-6.
10. BS-4449-1997/British Standard 4461-1978.
11. Clay-Arkansas-Geological Survey U.S.A. Geological survey Bulletin 351, 247p Braner J.C-1908
12. General Specification NHA – General Specification Ring Road. Package III, Lahore.
13. General and Particular Specification (Special Provision) Volume III & II MTBS-NESPAK.
14. Asphalt Concrete, The best surface for parking areas-Asphalt Institute.
15. Cracking of Bituminous Surfacing in High Ambient Temperature by J. Mc Elvaney, Highway and Transportation July 1991.
16. Hand Book for Cement Engineer – Abdul Hamid
    Cement Technologist and Consultant.
17. AASHTO – Designation MP-8.
18. Seminar Paper on stone matrix asphalt for Highways and heavy traffic roads in Pakistan by Mr. Shamim Zafar. Standard AASHTO – MP – 8)
19. Specification- Civil work- New Khanki Barrage Volume II
20. Gen-Specification- Ring Rd. Lahore package-3
21. Volumes of AASHTO, ASTM standards, American concrete Institute (ACI). Asphalt Institute volumes.
22. Ground granulated Blast furnace slag. Technical Bulletin TB 0102. W.R.Grace U.S.A.

www.ingramcontent.com/pod-product-compliance
Lightning Source LLC
Chambersburg PA
CBHW050528240926
51846CB00003B/735
9789690027283